T5-CQA-880

CHEMICAL ENGINEERING

at the

UNIVERSITY OF ARKANSAS

CHEMICAL ENGINEERING

at the

UNIVERSITY OF ARKANSAS

A Centennial History, 1902–2002

by

Michael S. Martin

Phoenix International, Inc.
Fayetteville
2002

Manufactured in the United States of America

06 05 04 03 02 5 4 3 2 1

Designer: John Coghlan

∞ The paper used in this publication meets the minimum requirements of the American National Standard for Permanence of Paper for Printed Library Materials Z39.48-1984.

Library of Congress Cataloging-in-Publication Data

Martin, Michael S.

Chemical engineering at the University of Arkansas : a centennial history, 1902-2002 / by Michael S. Martin.

p. cm.

Includes bibliographical references and index.

ISBN 0-9713470-0-X (hardcover)

1. University of Arkansas, Fayetteville. Dept. of Chemical Engineering—History. 2. Chemical engineering—Study and teaching (Higher)—Arkansas—Fayetteville—History. I. Title.

TP172.A7 M37 2002

660'.071'176714—dc21

2002003661

Contents

Acknowledgments

This book would not have been possible without the support of the University of Arkansas Department of Chemical Engineering faculty, alumni, and students. Although it is impossible to single out every person among that group who made a positive impact on the work, two faculty members stand out. Professor Buddy Babcock, who was chair of the department at the time, got plans for the centennial history underway in the mid-1990s; without his encouragement and determination, this book would not exist. Colonel William A. Myers, associate department head, ably imparted his knowledge, both first-hand and learned, about the department's history. His enthusiasm and support for the project has been an inspiration.

I would be remiss if I did not thank the faculty and alumni who responded to questionnaires, granted interviews, and answered questions over the year or so that I worked on this project. Everybody associated with the department has been cooperative and helpful; I suspect that this is a reflection of the department's influence on them. I owe particular debts to Leigh Downey and Janet Bowlin of the department, Andrea Cottrell of the University of Arkansas Special Collections, and John Coghlan of Phoenix International, Inc. Although all have participated in the creation of this book, any errors within it are mine exclusively.

Finally, I must thank my wife, Amy, for her unending patience and support. This book is dedicated to all U of A chemical engineering alumni and to my dad, an engineer who helped raise a historian.

Preface

When the word spread that Michael Martin was writing a history of the Department of Chemical Engineering at the University of Arkansas, I was delighted. And when he allowed me to read several chapters it brought back sixty years of fond memories. He was able to portray the dynamic leadership of Dr. Barker, who ushered in the modern era of chemical engineering in Arkansas; the evolving nature of the curriculum; the progressive changes under Bocquet, Couper, Gaddy, and Babcock; and, of course, the dramatic improvements in facilities, from the "Barn" to Bell Engineering Center. These things didn't happen by accident and would have been impossible without the cooperation and leadership of Deans Stocker, Branigan, Heiple, Halligan, and Schmitt.

From its beginning in the Department of Chemistry, with one or perhaps two chemical engineers on staff, the chemical engineering curriculum had an ample number of engineering courses as shown in this document. Many of the faculty (Hale, Dyer, Porter, Humphreys, Wertheim in chemistry, and Wray, Paddock, Kirby in engineering) were role models for a number of future instructors. To this list can be added the names of Adkisson and Richardson (mathematics), Sharrah and Hamm (physics), and others whose classes I did not have the pleasure of attending.

Doctor Barker was fond of saying that a chemical engineering degree would serve an individual well in any profession, and certainly this has been true. If one were to scan the resumés of graduates, he would find chief engineers, directors of divisions, vice presidents, presidents, CEOs of major national and international corporations, professors in prestigious universities, consultants, and owners of thriving businesses. Some former students in the department have, as Barker predicted, been highly successful as doctors, dentists, financial consultants, architects, and a host of other professions not directly related to chemical engineering.

Perhaps the greatest joy of teaching is to see young people develop in their chosen field and to monitor their activities after they were graduated. Certainly, the university has been blessed with bright, dedicated students. It has been said that good students make teachers look good. If this is true, we have, for many years, had one of the best faculties in the country!

This history should be a part of the library of every graduate of the Department of Chemical Engineering of the University of Arkansas.

PREFACE

Michael Martin has done an excellent job of both researching and writing. In writing a preface, it is tempting to reminisce about the "good old days," but we will leave that until another day.

Charles "Epi" Oxford
Emeritus Professor of Chemical Engineering
BSChE, '44

Introduction

Chemical engineering is a distinctive discipline. It is neither completely chemistry nor completely engineering. Instead, it is a unique combination of the two, separate from chemistry and the other natural sciences and also from the other engineering fields. But it has not always been so, especially in terms of chemical engineering education.

During the early twentieth century, chemical engineering educational programs in the United States tended to fall on one side of the fence or the other; they leaned heavily toward chemistry or toward engineering, but they usually did not combine the two into a separate and unique entity. Indeed, the process of creating educational programs that were neither chemistry nor some other form of engineering but were uniquely chemical engineering sometimes took decades.

This process occurred at colleges and universities around the United States, and the University of Arkansas (U of A) was no different. When the U of A instituted its first chemical engineering curriculum in 1902, the coursework leaned heavily toward the chemistry side of the spectrum; it did, however, include some work in various engineering fields. Over the next few decades, chemistry remained supreme for young chemical engineering students. It was not until the 1932–1933 school year that what might be termed the first true chemical engineering course, Elements of Chemical Engineering, was offered, and it took until 1936 for an instructor with extensive experience in the field to arrive at the U of A (he, however, had been trained in college as a chemist, not a chemical engineer). The first instructor with actual chemical engineering degrees, Stuart McLain, arrived in 1938, and the chemical engineering program separated from the chemistry department and became a distinct Department of Chemical Engineering in the College of Engineering in 1945.

McLain's arrival marked the beginning of a transitional period for the U of A chemical engineering program. Over the next decade or so, the department existed in a state of flux as World War II took precedence over faculty and student continuity. In 1948, Colonel Maurice Barker took over the program as departmental chair and instituted a period of long-term stability and continuity that continues even today. Barker's arrival marks a definite turning point for the program. Up to 1948, the program had been searching for an identity, and, thanks to Colonel Barker and succeeding departmental chairs, it found one: first and foremost, it is a department dedicated to preparing undergraduates for professional careers; more gradually it became a center for advanced chemical engineering research by faculty, graduate students, and undergraduates.

The chemical engineering program has exhibited a remarkable continuity in terms of the people involved with it. Faculty members taught, and continue to teach, students decades after they arrived at the U of A. After the creation of master's and doctoral program's, former undergraduates returned to the Fayetteville campus to increase their knowledge—quite often, former students of the program returned to the university as members of the faculty. This continuity of students and faculty resulted in the long-term impact of individuals on chemical engineering at the U of A—decisions made in the first decade of the twentieth century reverbrated through the 1940s; plans and hirings of the 1960s continue to be important today.

Outside of the transitional period (1938–1948) of its history, the U of A chemical engineering program has been guided by strong faculty and administrative personalities. For this reason, most of the chapters in this book are titled according to the head of the program during specific time periods. Individual biographical sketches have been included at the conclusion of each chapter for faculty members of particular importance. Not all faculty members have been sketched, but each chapter also contains a list of faculty members from chemistry, engineering, and chemical engineering who would have significantly impacted students of their particular era.

Each chapter also includes discussions of curricular changes, the physical environment, and extracurricular activities. Because research gradually became of central importance to the chemical engineering program, sections have been included on this subject in later chapters. Discussions of student life have been included, as well, but they have been especially problematic. Due to the demanding nature of the program, chemical engineering students seem to have less free time than students in some other fields, and, as the twentieth century wore on, student life at the U of A in general became much more diverse and much harder to pin down for the historian. Whereas in the early decades of the twentieth century students had few options for outside-the-classroom diversions, later students had increasingly more and different opportunities for fun and recreation.

INTRODUCTION

Before tracing the history of chemical engineering at the University of Arkansas, a brief overview of the university's chemistry and engineering programs of the late nineteenth century is useful.

The Arkansas Legislature established Arkansas Industrial University (AIU, today's University of Arkansas) in 1871 as the state's land-grant university. Professor T. L. Thompson was named chair of theoretical and applied chemistry and taught the first chemistry classes at AIU from 1873 until his death in 1875. F. L. Harvey succeeded Thompson and hired Cuthbert Powell Conrad as adjunct professor in 1879. Two years later, in 1881, the university's science departments were subdivided: Harvey took control of biology and geology, and Conrad took over chemistry and physics. Professor Conrad was determined to raise the level of scholarship at the university, and his efforts earned the enmity of many a student in those early days. He, along with every other faculty member at the university, lost his position in a general overhaul of AIU in 1885. Conrad's replacement, George D. Purinton, was named professor of chemistry, mineralogy, geology, and biology. Purinton left when the responsibilities of superintendent of agriculture were added to his position in 1887. Professor Albert Ernest Menke replaced Purinton. Menke remained at the university (its named changed to the University of Arkansas in 1899) until 1902, when A. M. Muckenfuss, the founder of chemical engineering at the University of Arkansas, arrived.

Arkansas Industrial University offered its first engineering courses during the 1873–1874 school year. These early courses focused mostly on mechanical engineering with a heavy dose of science, but by 1878, students could receive the degree of C.E. (civil engineer). In 1880, coursework leading to the degree of M.E. (mining engineer) was added, but it was dropped in a general reorganization of engineering offerings in 1886. From 1886, students could earn degrees in either mechanical or civil engineering; electrical engineering was added in 1891.

From 1872 to 1875, General N. B. Pearce, professor of mathematics and engineering, taught the very first engineering courses at the U of A. Pearce's replacement, Colonel O. C. Gray, served as professor of mathematics and civil engineering until 1879. In 1880, the university created the chair of applied mathematics and engineering; its first occupant,

J. B. Gordon served until 1884, when J. M. Whitham replaced him. Whitham served until 1891. In that year the university began to differentiate the separate fields of engineering: C. V. Kerr was hired as superintendent of mechanic arts and professor of engineering; G. C. Schoff became adjunct professor of civil engineering; H. B. Smith was named instructor of electrical engineering. In 1893, J. J. Knock took control of the civil engineering classes. The following year, William Nathan Gladson began teaching the electrical engineering coursework, and, in 1896, G. M. Peek replaced Kerr in mechanical engineering.

In 1902, the University of Arkansas created the College of Engineering. All the older engineering fields came under the new college's jurisdiction. One new field, however, did not. Chemical engineering, offered for the first time that year, would have to wait decades to become a part of the College of Engineering.

Chapter One

Foundations, 1902–1918

Chemical engineering education in the United States developed in a rather haphazard fashion. The profession lacked a definite relationship with other disciplines, whether in the sciences or engineering, and this ambiguity raised questions as to whether educators trained students to be chemists, engineers, or some unique mix of the two. Of course, the profession finally settled upon the latter distinction, but during these early years a major difficulty developed as to whether chemistry or engineering departments should oversee chemical engineering education.

Many schools followed the lead of the Massachusetts Institute of Technology (MIT), where the first chemical engineering program, Course X, had been created in 1888, and designed their chemical engineering curricula as offshoots of chemistry departments. At other institutions, chemical engineering developed as "free-standing units" with no links to other departments; a much smaller number of chemical engineering programs originated in engineering departments. Chemical engineering at the University of Arkansas followed the path marked by MIT and developed under the aegis of the Department of Chemistry.

In 1902, the U of A's Board of Trustees, at the urging of Professor of Chemistry and Physics A. M. Muckenfuss, approved the creation of a new degree curriculum in chemical engineering. The first mention in the university's catalog of the new bachelor of chemical engineering (B.Ch.E.) degree came in 1903, when the following course of study was prescribed:

1. Freshman Year: five hours of Mathematics (including Solid Geometry, Trigonometry, and Algebra), three hours of General Chemistry, two hours of Shop Work, two hours of drawing, and three hours of elementary German.
2. Sophomore Year: five hours of Analytical Geometry, four hours of Qualitative Analysis, two hours of Blowpipe Analysis, three hours of Elementary Physics, three hours of Elementary Machine Design, and two hours of ShopWork.
3. Junior Year: eight hours of Quantitative Analysis, three hours of Calculus, four hours of Theoretical Mechanics/Mechanics of Materials, three hours

of Metallurgy, three hours of Assaying, and two hours of Electrical Measurements.

4. Senior Year: three hours of Theoretical Chemistry, three hours of Electro-Chemistry, two hours of Gas Analysis, two hours of Water Analysis, two hours of Sanitary Engineering, three hours of Waterworks Engineering, three hours of Steam Machinery/Engines/Boilers, three hours of thesis work, and three elective hours.

None of the courses was labeled specifically as chemical engineering. The bulk of the coursework came from the chemistry department—chemical engineering students actually had to take more chemistry courses than chemistry majors—or from the Department of Mechanical Engineering.

In addition to the coursework leading to the bachelor's of chemical engineering degree, the 1903–1904 catalog notes the existence of a chemical engineering (Ch.E.) degree, similar to what might be called a graduate or professional degree today. This degree gave "additional preparation for those students who have finished an undergraduate course in engineering for some special line of work to which their previous study has led." Coursework for the Ch.E. degree comprised "one principal subject based on the course already pursued, and two secondary subjects, one or both of which should be closely related to the principal," and included no less than fifteen weekly hours of work for a year. Additionally, students were required to write a thesis. Graduates with a B.Ch.E. degree from the university could receive the Ch.E. degree if they had been "in successful practice of their profession for three years, and . . . [had] submitted a satisfactory thesis on a subject approved by the faculty." Although there are no records of any Ch.E. degrees being granted, it is interesting to note that, even at such an early date, the program's faculty and administration had created an option for what amounted to a graduate degree.

Professor Charles G. Carroll, who replaced Muckenfuss in 1905, instituted a number of changes to the chemical engineering curriculum. By 1905, Elementary English Composition had replaced Elementary German, Advanced English Composition had replaced Blowpipe Analysis, and the senior-year elective had been replaced with an Electrical Engineering Lab. In 1907, Business Law was added to the senior year, and the foreign language component was brought back (students could choose either German or French). By 1911, both French and German were required, and the Business Law requirement had changed to Engineering Law. By 1913, all engineering freshman were taking a similar course load of math, English, physics, and mechanical engineering courses (chemical engineering majors substituted Elementary Chemistry for mechanical engineering shop work, though). Also that year a senior-year elective replaced the Engineering Law requirement. More electives were added in 1915, as was a civil engineering course in 1916.

By the 1917–1918 school year, chemical engineering students at the U of A had to take the following coursework:

Freshman Year

Semester One:
1. College Algebra
2. Plane Trigonometry
3. Rhetoric & Composition
4. General Chemistry
5. General Shop Practice
6. Drawing
7. Military Art 1

Semester Two:
1. Solid Geometry
2. Analytic Geometry
3. Rhetoric & Composition
4. General Chemistry
5. General Shop Practice
6. Elem. Descriptive Geometry
7. Military Art 1

Sophomore Year

Semester One:
1. Qualitative Analysis
2. General Physics
3. Advanced Algebra
4. Differential & Integral Calculus
5. Military Art 2
6. Elective

Semester Two:
1. Quantitative Analysis
2. General Physics
3. Elective
4. Differential & Integral Calculus
5. Military Art 2

Junior Year

Semester One:
1. Organic Chemistry
2. Advanced Qualitative Analysis
3. German
4. Elective

Semester Two:
1. Organic Chemistry
2. Advanced Quantitative Analysis
3. German
4. Elective

Senior Year

Semester One:
1. Physical Chemistry
2. Water Analysis
3. Industrial Chemistry
4. Advanced Inorganic Chemistry
5. Elective

Semester Two:
1. Electro-Analysis
2. Physical Chemistry
3. Advanced Inorganic Chemistry
4. Elective

Note that, even by 1917, the U of A still offered no courses labeled specifically as "chemical engineering." The 1917–1918 catalog noted that "all electives must be chosen with the consent of the head of the department of Chemistry and Dean of the College of Engineering. Three hours, in the Sophomore year, and six hours, in each the Junior and Senior years, must be elected from the college of engineering."

For the period 1902–1918, students enrolled in chemical engineering at the U of A took most of their classes from professors in the Department of Chemistry and the College of Engineering. Because chemical engineering fell under the auspices of chemistry, the following list contains all professors and instructors within that department. Also included are the professors of engineering any chemical engineering student would have encountered.

CHEMISTRY

1. A. M. Muckenfuss. Professor of Chemistry and Physics (1902–1904). Professor of Chemistry and Head of the Department of Chemistry (1904–1905). B.A., M.A., Wofford College; Ph.D., The Johns Hopkins University.

2. Louis Henry Rose. (1902–1904).

3. Charles G. Carroll. Professor of Chemistry and Head of the Department of Chemistry (1905–1916). A.B., A.M., Southwestern University; Ph.D., The Johns Hopkins University.

4. Hugh E. Morrow. Adjunct Professor of Chemistry (1904–1907); Associate Professor of Chemistry (1907–1920). B.S.A., University of Arkansas.

5. James Samuel Guy. Professor of Chemistry and Head of the Department of Chemistry (1916–1918). B.S., M.A., Davidson College; Ph.D., The Johns Hopkins University.

6. Harry Elwyn Sturgeon. Assistant in Chemistry (1916); Instructor in Chemistry (1917–1918). B.A., Cooper College; M.S., Purdue University.

7. Dana Porter Weld. Assistant in Chemistry (1917). B.S.C., University of Arkansas.

ENGINEERING

1. Burton Neill Wilson. Mechanical Engineering. B.Sc.M.E., Georgia School of Technology; M.E., University of Michigan; M.M.E., Cornell University.

2. Brainerd Mitchell. Mechanical Engineering. B.M.E., University of Arkansas.

3. William Nathan Gladson. Electrical Engineering. B.M.E., E.E., Iowa State College; Ph.D., McLemorsville College. Gladson also served as Dean of the College of Engineering from 1913 to 1936.

4. Lee Sedwick Olney. Electrical Engineering. B.E.E., University of Arkansas.

5. Julius James Knoch. Civil Engineering. B.S., M.S., Grove City College; C.E., Cornell University.

6. Virgil Proctor Knott. Civil Engineering. B.C.E., University of Arkansas.

The earliest students of chemical engineering at the University of Arkansas knew a very different campus than those of today. For the most part, their dormitories, classrooms, and laboratories have all disappeared. One building, however, remains: Old Main, even more then than now, was the centerpiece of campus. In addition to classrooms and faculty offices, University Hall, as students in the early twentieth century called the building, housed the school's library, gymnasium, and bookstore.

The very first chemical engineering students took most of their chemistry classes in Science Hall, a two-story brick building constructed in 1893

The Campus, 1907, from Mount Nord. (*Cardinal,* 1907, 6)

and located on the south side of campus. Professor A. M. Muckenfuss's office was located on the first floor, along with lecture rooms, a storeroom, and physics laboratories. The second floor of the fifty-by-ninety-foot building contained laboratories for general chemistry, qualitative analysis, and quantitative analysis; a private laboratory; a chemical supply storeroom; and a weighing room. According to the 1902–1903 catalog, "the building is supplied with gas, water, steam heat, and with modern appliances for technical work." It held up to 150 students.

The general chemistry laboratory had space enough for one hundred students at desks that were supplied with gas and water. The qualitative and quantitative laboratories each could fit up to sixteen students and were arrayed with "the usual equipments." The weighing room contained "two of Becker's best analytical balances, besides a number of less accurate instruments suitable for weighing large quantities of chemicals." An assistant ran the storeroom and kept track of the supplies, including "the apparatus for preparing distilled water."

In 1905, the Arkansas General Assembly provided funds to build six new buildings on the university's campus, including a new Chemistry Building. The new buildings would be up to "the latest standards," and, according to the 1905–1906 catalog, "all of them are fire-proof construction, with brick walls, slate roofs, stone foundations, and granite trimmings." Of the six buildings, three still stand on the campus today—the Chemistry Building (the Student Development Center), the Agriculture

Science Hall. (*Catalog,* 1902-1903, v)

Building (the Agriculture Annex), and the Young Women's Dormitory (Ella Carnall Hall).

The Chemistry Building, designed by the Thompson and Gates firm of Little Rock for a cost of eighteen thousand dollars, was constructed just north of Old Main. The first floor of the two-story building housed qualitative and quantitative analysis laboratories, organic and physical chemistry laboratories, faculty offices and laboratory space, a balance room, a library, and a room for analyzing fertilizers. On the second floor, young chemical engineers and chemists found a general lecture room for one hundred and fifty students and the general chemistry laboratory. The basement contained storage space, furnaces, and a gas machine. According to the 1906–1907 catalog, "the various laboratories are well provided with work-tables, sinks, hoods, water, and gas."

Engineering classes met in the Engineering Building, a three-story brick building constructed in 1904 for twenty-five thousand dollars. The building, roughly one hundred fifty feet by sixty feet, housed the departments of electrical, civil, and mechanical engineering and was divided by hallways bisecting its length and width. It included offices, lecture halls, and laboratories, plus an assembly room, the engineering library, and a reading room.

Chemistry Building. (Mullins Library Special Collections, photo collection #1023)

Of the students attending classes in the Chemistry Building and Engineering Hall between 1902 and 1918, only two graduated with degrees in chemical engineering. Leslie Clair Hughes of Fayetteville and Robert F. Hyatt of Monticello became the first students to graduate from the University of Arkansas with B.Ch.E.s in 1907. Hughes had been vice

Engineering Building. (Mullins Library Special Collections, photo collection #1087)

president of the senior class and played trumpet in the Cadet Band. Hyatt had been athletically inclined; he starred as captain and pitcher for the baseball team and halfback for the football team. Hyatt later had a long military career; Hughes worked in the chemical industries. The next graduate in chemical engineering, Robert Renic Logan, did not receive his diploma until 1919.

Leslie Clair Hughes. (*Cardinal,* 1907, 24)

Although only two students graduated with chemical engineering degrees between 1902 and 1918, the program had students throughout the period. These earliest chemical engineering students rarely had field-specific extracurricular opportunities, but they did take part in activities with their fellow chemists and engineers.

Engineers Day was especially important for future engineers. First held at the U of A on March 17, 1909, the day was intended to honor St. Patrick, patron saint of engineers. Early Engineers Days included many of the hallmarks of later ones, according to one account: "The whole day is given over to various activities of the engineers. This year the program was a grand parade over town, engineering ceremonies in front of University Hall [Old Main], including the kissing of the Blarney Stone by the seniors and their initiation into the Order of the Knights of St. Patrick; exhibits all the afternoon and a dance and entertainment at night. The *University Weekly,* under a special

Robert F. Hyatt. (*Cardinal,* 1907, 25)

staff of engineers, issued an engineer's edition. The day was a great advertisement for the U. of A. engineering departments."

> Description of the first Engineers' Day from "A Brief History of Engineers' Day Celebrations," by L. Gale Huggins (1921):
>
> The morning of St Patrick's day the Engineers all assembled in the chapel where St. Patrick outlined the plans for the day. A parade was then formed and after making the rounds of the main building they paraded past Carnall hall, through Schuler and up Dickson street to Block and thence to the square. Upon arriving up town they stopped on the west side of the square and had a picture made. At 10:30 the laboratories were opened to the public. Many interesting and puzzeling [sic] stunts were on display in the various labs.
>
> The students of the Civil Engineering Department had their tents up on campus. The tents housed a number of side shows, to which everybody was welcome, no admission being charged.
>
> One of the features of the day was the knighting of the seniors who were to graduate. A Blarney Stone, cast from concrete was used in this ceremony. St. Patrick conducted the Knighting ceremony and each senior kissed the Blarney Stone, thereupon becoming a Knight of St. Patrick. This knighting of seniors has always been one of the features of Engineers Day.
>
> C. M. Moreland was St. Patrick at the first Engineers Day celebration at the University.
>
> At 12:30 a big banquet was held on the campus south east of the Engineering Building. Needless to say all Engineers were on hand at this time and [sic] as there were plenty of good things to eat. All enjoyed themselves immensely. The chief attraction for the afternoon consisted of a baseball game played between a team composed of EE's and ME's against the CE's and the ME's (Mining Engineers.)
>
> The Engineers Dance, always one of the biggest of the school year, was no exception to the rule this time. The drawing tables and stools were cleared out of all of the rooms on the first and second floors and the dance was held in the Engineering Building. The orchestra was located in the center of the first floor hall and the halls and all the rooms on both floors were used for dancing.

Upon his arrival at the U of A in 1905, Professor Carroll created an informal Journal Club, in which chemical engineers could participate. The club, which included chemistry instructors, Agricultural Experiment

Early Engineers' Day, March 17, 1915. (*Razorback,* 1915)

Station chemists, and advanced chemistry students, met every other week to discuss and report on recent developments in their field. Following Carroll's death, Professor James S. Guy encouraged the club to reorganize itself "so that its work might be carried on under the direction of officers elected by the membership." Thus reconfigured, the members renamed it the Chemical Club and held bimonthly meetings to discuss contemporary chemical interests. Out of the Chemical Club grew Gamma Chi, which was founded in 1918 and was the precursor to the Alpha Sigma chapter of Alpha Chi Sigma.

Beginning in the fall of 1915, chemical engineering students also could participate in the Chemical Engineering Society, which was open to all students of the field. The organization elected W. D. Merrill as its first president. According to the 1916 yearbook, "While young among the other societies, this society means to make itself felt in engineering circles."

Besides the extracurricular activities that they shared with chemistry and engineering students, young chemical engineers of the early twentieth century shared a common experience of university life with all students on the Fayetteville campus. Looking back from the beginning of the twenty-first century, today's observer would hardly recognize campus life of the early twentieth.

In order to enroll in the 1902 freshman class, all students had to report directly to the university's president and present "certificates of honorable discharge from the school last attended, or furnish other testimonials of good moral character." Unless they had graduated from an accredited preparatory school or from a reputable college or university, the students had to pass an entrance exam. Even then, chemical engineering students would not be admitted without meeting the following minimum preparatory requirements: eight credits of English; five credits of Algebra; four credits of plane geometry; three credits of United States history; and four credits of Latin, Greek, French, German, physical geography, physiology, botany, zoology, physics, chemistry, English history, civil government, bookkeeping, freehand drawing, or shop work.

Once they were admitted, students lived a very different life from those of today. In 1902, male students lived on campus in one of the dormitories for boys (Hill Hall and Buchanan Hall) or boarded off campus. Dormitory rooms came free of charge but lacked any amenities, including furniture and lighting. In-state students received first choice for dorm rooms. Male students living off campus could board only at places personally approved by the president of the university, and they could not move without the president's consent.

Women who attended the university had to live off campus, as there was no dormitory for them in 1902. According to the university's catalog of 1902–1903, "all necessary assistance is rendered them in finding homes in private families in the town. Parents, therefore, who send a daughter to the University, should, place her under the control of the family with

Chemical Engineering

Faculty.

DR. C. G. CARROLL   PROF. H. E. MORROW

Roll.

Dean, T. O.	Merrill, W. D.
Henderson, E. L.	Ramsey, W. L.
McDaniel, V. B.	Thompson, O. C.

The chemical engineering course in the University of Arkansas has not as many students as the other engineering courses, but is nevertheless one of the most thorough courses offered. Seventy-two hours are required for graduation, as compared with sixty-seven in the other departments. The regular chemical engineering course embraces a thorough training in chemistry and extensive courses in mechanical and electrical engineering. The graduates are trained for analytical work, physical and industrial chemistry and other occupations.

1916 Chemical Engineering Students and Faculty. (*Razorback,* 1916, “Engineers Section,” unnumbered page)

whom she boards, subject to the general supervision of the president of the University."

Relationships between men and women were governed by strict rules, as noted in the 1907–1908 *Catalog:*

> I. Young ladies and young gentlemen are not allowed to board at the same place.
> II. Young ladies are not allowed to change their boarding places without permission from the Dean of Women.
> III. Callers may be entertained only on Friday and Saturday evenings and also on Sunday evening when a young man desires to accompany a young lady to church.
> IV. Callers are expected to leave at 10 o'clock, p.m.
> V. Young ladies may go out only on Friday and Saturday evenings. This regulation may be suspended for lectures or other high class entertainments.

In 1905, under the same legislation that created the Chemistry Building, the Arkansas General Assembly provided funding for a third men's dormitory and the first women's. The men's dorm, known as Gray Hall and located to the west of Old Main, housed 136 young men, and was designed "to provide as many bed rooms as possible with every comfort and convenience patterned after the U.S. Army barracks." The women's dorm, Ella Carnall Hall, which still stands on the northeast corner of campus, housed 110 students and was "designed to be complete within itself, having its own toilet and bath rooms, dining rooms, kitchen, and an independent steam heating plant."

There were no Greek letter organizations in 1902. The Arkansas General Assembly had abolished them in 1901. However, a number of informal clubs—such as the Success Club for men, the Eclectic Club for women, and the co-ed A.O.T. Club (motto: "To eat, drink, and have all the fun that can be crowded into a single evening.")—arose to replace them. The anti-fraternity law was never enforced to any great degree (it was repealed in 1929), and by 1906 sororities and fraternities were noted in the university catalog. The 1910–1911 catalog stated that students could not be initiated into fraternities or sororities until they had completed fifteen hours of coursework in one term, adding, "if any fraternity shall permit its members to drink wine, whiskey, beer, or other intoxicants, in its chapter house or meeting place, or allow such liquors to be kept or stored there, or shall permit any gambling or other violation of law therein, or

Buchanan Hall. (*Cardinal,* 1914, unnumbered page)

Hill Hall. (Reynolds and Thomas, 128)

Carnall Hall. (Reynolds and Thomas, 184)

Gray Hall. (Mullins Library Special Collections, photo collection #1126b)

shall keep a disorderly house or place, such fraternity shall be cited for trial before the faculty, and upon proof being adduced establishing any of the above-mentioned offenses, such fraternity shall not be allowed to exist longer under its own name or under any other form or name in the University of Arkansas."

All male students, unless physically unable, were required to participate in military drill. Drill occurred three days a week during the months of October, November, March, and April, and included "infantry drill, target practice, camping, guard duty, and various other exercises." Drill was expected to develop male students mentally and physically: "It contains a course of gymnastic exercises for the development and improvement of the arms, chest, legs, hands, and feet. . . . The necessity of being alert, listening for each word of command, and acting promptly on it, quickens the wit and cultivates the habit of fixing the attention and concentrating the thoughts." All participants in drill were required to wear a gray uniform with brass buttons and black trim. In lieu of participation in drill, women were required to take an extra hour of coursework. At the university's annual commencement exercises the cadets participated in group and individual competitions for prizes.

Military drill. (*Cardinal,* 1907, 152)

All university students were expected to be practicing Christians. Indeed, the faculty and administration of the University of Arkansas during the early twentieth century placed great emphasis on religious instruction and duty. To that end, all students were required to attend university-sponsored Christian training sessions; in 1902, this meant daily religious exercises held in the university chapel. Students also received religious instruction, in addition to social opportunities, through the Young Men's and Young Women's Christian Associations (YMCA and YWCA).

In 1902, there were three literary societies on campus—two for men only, the Garland and Periclean Societies, and one for both men and women, the Mathetian Society. Literary societies had existed on the campus since 1872, only a year after the university's founding, and they provided both social and intellectual stimulation for students. According to U of A historian Robert Leflar, "debates, written essays and orations were the staples of literary programs, for the men. The women's societies put less emphasis on debate and oratory, more on dramatic readings and original writings of various kinds, including essays and poetry." Despite their popularity with students in the first decade of the century, the literary societies gradually died out.

"Pennant Day" also offered an enjoyable, though occasionally dangerous, outlet. This spring holiday began in March 1903, when the junior and senior classes were allowed a day off. The seniors raised their class pennant on the north tower of Old Main and dared the juniors to take it down; a melee ensued as the juniors attempted to take the flag and the seniors tried to repel them. The brawl became a yearly tradition, according to the 1932 *Razorback:* "Black eyes, shattered noses, and an occasional broken arm were always visible following the annual 'knock-down, drag out.' This tradition, unfortunately, was culminated when several class members fell or were thrown from the top of the tower. Much to the chagrin, perhaps, of their opponents they were caught and saved from serious injury by an open awning several floors below." As might be expected, the faculty ordered the fighting stopped but allowed the holiday to remain. From this point on, "in the morning they attend chapel and make the hall ring with yells and class songs. Afterwards for an hour or more they promenade the corridors and with yells, songs and divers noises they annoy the professors, who are valiantly struggling to impart knowledge to the absent-minded lower classmen. Next the juniors and seniors repair to one of the literary society halls where a short joint program is given." Following this, the classes would head to the front of Old Main, where they planted a senior tree and laid the concrete blocks that today make up the earliest sections of Senior Walk. On Pennant Day afternoon, the two classes played baseball and then, in the evening, held a joint banquet.

The first decades of the twentieth century marked the infancy of U of A athletics. The first football team had been coached by future university president John Clinton Futrall in 1894, but, through 1907, the team lacked adequate coaching and only occasionally won against teams of equal talent. In 1908, however, the university hired Hugo Bezdek as football coach and director of physical culture and athletics. Bezdek had played for and studied under legendary coach Amos Alonzo Stagg at the University of Chicago, and he brought Arkansas football its first period of glory. The 1909 squad became one of only two Arkansas football teams to ever complete a season undefeated; the 1910 team only lost one game. Perhaps even more important, Bezdek anointed the university's athletic teams as "Razorbacks." When Bezdek left to take over the University of Pittsburgh's team in 1912, the U of A football team returned to the level of mediocrity it had exhibited before his arrival.

Arkansas v. Oklahoma, 1909. (Reynolds and Thomas, 324)

Although baseball as an organized sport had been played since 1897, the Arkansas baseball teams of the 1902–1917 period never reached the level Bezdek attained with football. The team did post respectable records, however, with the best coming in 1908 (18–5) and 1911 (19–5).

Basketball was strictly a women's sport in the early 1900s. The first teams at Arkansas were created in 1904, but the first intercollegiate contests were not held until 1906. By 1909, the team had achieved enough prominence "to play Mississippi State College for Women for the 'championship of the South.'" The Arkansas team beat their hosts, 12–9, and then accepted an invitation to play the University of Illinois for a "national championship." The game never came off, however, because U of A president John N. Tillman and Coach Bezdek, claiming that basketball was being overemphasized at the expense of the young women's studies, completely shut down the program.

Although the earliest athletics contests were held in front of Old Main, by the early twentieth century the university had created an east-west field on what was then the southwestern corner of the campus (at the current site of the Fine Arts Center and the Arkansas Union). There, spectators could view football and baseball games from a covered grandstand. In 1914, Arkansas became a charter member of the Southwestern Athletic Conference, creating a tie that lasted for three quarters of a century.

The town of Fayetteville grew physically and economically during the early twentieth century, and it slowly began offering outside diversions for university students. Students with a cultural bent could attend performances at the Knights of Pythias Opera House (also known as the Ozark Theater), or they could gather on the square (before the "old" post office building was erected) for performances by Frank Barr and his musicians. Barr built the first motion picture theater, the Lyric Movie House, around 1911 on the corner of Meadow Street and North Block Avenue. Students

Center Street, ca. 1910. (Courtesy Washington County Historical Society / Shiloh Museum of Ozark History)

wishing to dine out could get a twenty-five-cent meal at Tom Stoke's Frisco Café. The annual Washington County Fair, held on the site of today's U of A intramural fields, provided diversions ranging from livestock shows to foot races. After about 1907, automobiles began to be seen in the town; leisure riding, however, was not often an option on the city's mostly unpaved roads.

College Avenue, 1907. (Courtesy Washington County Historical Society / Shiloh Museum of Ozark History)

All of these activities—social, cultural, religious, athletic—provided valuable outlets for the U of A's students of the early twentieth century. Chemical engineering students were no different. As the next few decades passed, an almost constantly increasing number of chemical engineering students would continue to take part in the life of the university.

By the time Professor of Chemistry J. Sam Guy left the University of Arkansas in 1918, the stage had been set for the gradual, yet constant, development of the chemical engineering program. From the seeds planted by Professor Muckenfuss and nurtured by Professors Carroll, Morrow, and Guy, the chemical engineering program began to mature under the watchful eye of Professor Harrison Hale over the next two decades.

Biographical Sketches, 1902–1918

Antony Moultrie Muckenfuss created the chemical engineering program at the University of Arkansas. Muckenfuss, born in 1869 in Charleston, South Carolina, received a bachelor of arts degree (1889) and master of arts (1890) from Wofford College. After brief tenures as principal of a South Carolina high school and as a student at the University of Virginia, he entered the Johns Hopkins University in 1894 and earned his Ph.D. there the following year. His postdoctoral studies included stints in Berlin, Germany, and at the University of Chicago. Muckenfuss came to the University of Arkansas from Millsaps College of Mississippi in 1902. He served as professor of chemistry and physics from 1902 to 1904, then professor of chemistry from 1904 to 1905. In 1905 he returned to Millsaps.

It was during his studies in Germany that Muckenfuss first came into direct contact with practices of applied industrial chemistry. The German chemical industries at the time far outpaced their closest national rivals, and chemists in Germany focused more on the practical uses of chemistry while downplaying purely theoretical forms of the science. Muckenfuss's German experience had a profound impact on him, and, combined with his experience as a chemist for the Lowe Paint Company of Ohio, impressed him as to the importance of applying chemical knowledge to real-world applications. Thus, when he

A. M. Muckenfuss. (Reynolds and Thomas, 498)

arrived at the University of Arkansas, one of his primary goals "was to persuade [the] University's board and administration to have a curriculum in chemical engineering."

Besides overseeing the creation of the first chemical engineering curriculum at the University of Arkansas, Muckenfuss taught Elementary Chemistry, General Inorganic Chemistry, Qualitative Analysis, Organic Chemistry, Quantitative Analysis, Agricultural and Food Analysis, Gas Analysis, and Water Analysis.

Charles Geiger Carroll replaced A. M. Muckenfuss as professor of chemistry in 1905. Born the son of a Methodist minister near Ashland, Kentucky, in 1875, Carroll grew up in West Virginia and Colorado. His early education included home-schooling, studies at the Pueblo College Institute and Central High School of Pueblo, Colorado, and in the University of Denver's preparatory department. He entered the University of Denver's freshman class at the age of fifteen, but his studies were interrupted by a move to Texas, where he taught and served as principal at two public schools. Finally, in 1895, he enrolled in Southwestern University (Texas), where he earned a bachelor of arts in 1896 and a master of arts in 1897.

Charles G. Carroll. (Reynolds and Thomas, 446)

For the next two years, Carroll served as instructor in Latin, English, French, chemistry, and physics at his alma mater. In 1898, he was named assistant professor of chemistry at Southwestern, and he channeled his energies into scientific studies. By 1905, Carroll had risen to the rank of professor of chemistry and also had earned his Ph.D. from the Johns Hopkins University.

In September 1905, Carroll became professor of chemistry and head of the Department of Chemistry at the University of Arkansas, where he served until his death in February 1916. His course offerings at the U of A included Inorganic Chemistry,

Qualitative Analysis, Quantitative Analysis I and II, Physical Chemical Assaying, Agricultural and Food Analysis, Water Analysis, Electro-Chemistry, History of Chemistry, Chemistry for Teachers, and Quantitative and Qualitative Spectral Analysis and Colorimetry.

In addition to his duties as professor of chemistry, which included supervising the chemical engineering curriculum and its students, Carroll served as secretary of the faculty and published numerous articles on chemical and nonchemical subjects. He also created the university's first official Glee Club (earlier vocal clubs had existed, but had not performed off campus as representatives of the university); Carroll took the Glee Club on yearly tours of the state. The 1909 *Cardinal* yearbook noted of Carroll: "He has had the responsibility of the Glee Club thrust upon him, but with his musical turn he will be able to acquit himself creditably. Tennis, though, is his long suit. He talks, hears, thinks, sees, smells, and tastes love games [but] didn't get a hit in the last year's Senior-Faculty base ball game."

Carroll died on February 23, 1916, after a lengthy battle with lepto meningitis. The *Fayetteville Weekly Democrat* noted Carroll's far-flung expertise in its obituary: "[He] was author of many publications on technical subjects which have gained him considerable prominence and which have been embodied in text books used extensively in this country as well as foreign countries. His research work along scientific lines has been almost phenomenal when his broad gauge of activities and his comparative youth are considered. He was also joint author of a text-book on "French Lyric Poets of the Post-Romantic Periods," and at spare times often afforded himself recreation by composing music." Carroll's replacement, Professor James Samuel Guy, served as head of the chemistry department until the end of the 1917–1918 school year.

Hugh Ellis Morrow, a native of Washington County, Arkansas, served as adjunct, then associate professor of chemistry under both Muckenfuss and Carroll. Born in 1882, Morrow received his B.S.A. degree from the University of Arkansas in 1904 and was named adjunct professor that same year. He took graduate work at the University of Chicago, and, in 1906, received a promotion to associate professor.

Morrow shared responsibility with Professor Carroll for most of the lower-level chemistry courses the U of A offered during his tenure. Among the classes he taught were Qualitative Analysis, Organic Chemistry, and Elementary Chemistry I and II.

Morrow remained at the U of A until 1920. By the mid-1920s, he was on the staff as an instructor in chemistry with the medical department of Cornell University. He died on June 16, 1929, in New York City.

Hugh Ellis Morrow. (Reynolds and Thomas, 497)

Chapter Two

The Hale Years, 1918–1938

When, on April 6, 1917, the United States entered an already raging Great War, American lives changed drastically. As the federal government struggled to equip and train hundreds of thousands of men for the military and forced the swing from a peace-time to a war economy, the nation's universities underwent dramatic changes. The Great War transformed the University of Arkansas and helped usher in the rapid growth of chemical engineering both nationally and at Arkansas.

The outbreak of war in 1914 interrupted a stream of chemicals that had flowed from Germany to the United States. By the time the United States began to engage in actual hostilities, it had become obvious that the nation would have to rely on domestic production for many of the chemicals it needed. To that end, new chemical plants would be needed, as would workers trained to design and manage those plants. Thus, there began an increase in not only chemical production but also in the number of chemical engineers graduating from the nation's universities.

This development became quite evident at the University of Arkansas. Robert Renic Logan, the university's third chemical engineering graduate, received his diploma in 1919, twelve years after Leslie Hughes and Robert Hyatt received the first two B.Ch.E.s. Over the next twenty years, under the guidance of Professor of Chemistry Harrison Hale, the chemical engineering program's student enrollment increased dramatically.

The postwar economic boom that lasted until 1929 mirrored a contemporaneous boom in the number of chemical engineering students at the University of Arkansas. By 1929, eighteen more graduates (by then earning the degree of B.S.Ch.E., not B.Ch.E.) had left the shadow of Old Main and embarked on various careers, some in chemical engineering, some not. Although this number might seem insignificant in comparison to the program's later output, compared to the first two decades of the twentieth century, the numbers represented a massive increase.

Perhaps surprisingly, the number of chemical engineering students continued to grow after the stock market crash of 1929. By 1938, the year before another European war broke out, fifty-nine more chemical engineers had graduated with their bachelor's degrees.

Throughout this twenty-year period, 1918–1938, four chemistry professors dominated the chemical engineering program—Harrison Hale, Edgar Wertheim, Allen Humphreys, and Lyman Porter. "Whatever we lacked in facilities," remembered one student of the 1920s, "we made up for in the faculty. That faculty was outstanding." The same student added: "I could never have asked for a better group of teachers, even though Wertheim scared me to death." Ralph Higbie, the first faculty member with true chemical engineering experience, also arrived during this period, but it is notable that neither he, nor any of the other faculty members of the time, was referred to as a professor of chemical engineering.

1. Harrison Hale. Professor of Chemistry and Head of the Chemistry Department (1918–1945); Emeritus Professor (1945–1966). B.A., Emory College; M.S., University of Chicago; Ph.D., University of Pennsylvania; LL.D, Drury College.

2. Edgar Wertheim. Associate Professor of Chemistry (1921–1929); Professor of Chemistry (1929–1951); Head of the Chemistry Department (1945–1951); Emeritus Professor (1951–1966). B.S., Northwestern University; B.P.E., Y.M.C.A. College, Chicago; M.S., University of Kansas; Ph.D., University of Chicago.

3. Allan Sparrow Humphreys. Instructor in Chemistry (1918–1921); Assistant Professor of Chemistry (1921–1941); Associate Professor (1941–1944); Dean of Men (1937–1945); Professor (1944–1955); Emeritus Professor (1955–1973). B.S., Drury College; M.S., University of Pennsylvania; B. Ped., Southwest Missouri State Teachers College.

4. Lyman Edward Porter. Instructor in Chemistry (1921–1927). Assistant Professor (1927–1937); Associate Professor (1938–1954); Professor (1954–1959); Emeritus Professor (1959–1976). B.A., M.A., Ph.D., Yale University.

5. Harvey McCormick Trimble. Instructor in Chemistry (1918–1922). B.S., University of Michigan.

6. Jacob Robert Meadow. Assistant in Chemistry (1925–1926). B.A., University of Arkansas.

7. Owen Lloyd Osburn. Instructor in Chemistry (1927–1929). B.S., Colorado Agricultural College.

8. Walter S. Dyer. Instructor in Chemistry (1929–1938); Assistant Professor (1938–1942); Associate Professor (1942–1945). B.S., University of Arkansas; M.S., Ph.D., University of Minnesota.

9. Paul Porter Sutton. Instructor in Chemistry (1936). Ph.D., The Johns Hopkins University.

10. Ralph W. Higbie. Instructor in Chemistry (1936–1938). B.S., M.S., Sc.D., University of Michigan.

11. C. Truman Steele. Instructor in Chemistry (1936–1937). B.S., Drury College; M.S., University of Arkansas.

12. William McMurray Dix. Instructor in Chemistry (1937–1938). B.S., Washington and Lee University; M.S., Ohio State University.

13. Warren H. Steinbach. Instructor in Chemistry (1937–1939); Assistant Professor (1939–1943). B.S., Hastings College; M.S., Ph.D., University of Nebraska.

These professors and instructors guided both chemistry and chemical engineering majors as they sought their degrees, and they fashioned curricula to fit the needs of their times.

World War I brought drastic changes to the University of Arkansas's academic calendar. In order to shorten the length of time required to receive a degree, the university's administration went to a three-term (fall, winter, and spring) school year. This system continued until 1925.

Students enrolling in chemical engineering in 1918 could expect the following degree plan to receive the B.Ch.E.:

Freshman Year (54 credit hours)

Chemistry 141, 142, 143: General Chemistry
English 141, 142, 143: Rhetoric and Composition
Mathematics 156, 157, 158: Plane Trigonometry, College Algebra, Solid and Analytic Geometry
Drawing 121, 122, 123: Mechanical Drawing
Mechanic Arts 121, 122, 123: General Shop
Military Art 111, 112, 113

Sophomore Year (54 credit hours)

Mathematics 251, 252, 256: Differential & Integral Calculus, Analytic Geometry
Physics 241, 242, 243: General Physics
Drawing 221, 222, 223: Mechanical Drawing
Chemistry 251, 254, 255: Qualitative, Quantitative, Advanced Quantitative
Military Art 221, 222, 223

Junior Year (54 credit hours)

Chemistry 354, 355 or 651, 652: Organic
Chemistry 359: Industrial
Civil Engineering 251: Surveying
Experimental Engr. 232, 225: Mechanical Laboratory, Engines and Boilers
Elec. Engineering 241, 211: Elements of Electrical Engineering, Lab
Electives (24 hours)

Senior Year (54 credit hours)

Chemistry 434, 435, 436: History of Chemistry, Advanced Inorganic
Chemistry 451, 452: Physical
Heat Power Engineering 341, 342, 343: Theoretical Mechanics
Elective (23 hours)

The following year, all engineering students were required to take a common freshman and sophomore curriculum. The coursework for the first two years included Physics 144, 145, 146 (Experimental Physics); English 141, 142, 143; Mathematics 155 (Solid Geometry), 157, 128

(Solid and Analytic Geometry), 139 (Advanced Algebra); Drawing 121, 122, 123; Military Art 111, 112, 113; Mathematics 251, 252, 256; Chemistry 257, 258, 259; Drawing 221, 222, 223; Civil Engineering 251; Experimental Engineering 225, 232; Electrical Engineering 221, 231 (Elements and Lab); Military Art 221, 222, 223. This common curriculum for freshman and sophomore engineering students remained in effect for the next twelve years. In 1934, the engineering college dropped the common sophomore year but retained the freshman.

The common freshman and sophomore curriculum allowed junior and senior chemical engineering students to focus almost solely on their chosen specialization. To that end, the degree plans required coursework of mostly chemistry. In 1919, for instance, junior chemical engineers took twenty-five hours of chemistry courses, twelve hours of Heat Power Engineering, and seventeen hours of electives. That year's seniors took twenty-four hours of chemistry and thirty hours of electives. The catalog included the following qualification regarding electives: "All electives must be chosen with the consent of the head of the Department of Chemistry and the Dean of the College of Engineering. Of these electives 12 hours must be chosen from other courses in chemistry and at least 9 hours in English or a foreign language."

In 1921, fifteen hours of electrical engineering courses replaced half of the senior year's electives. Two years later, twelve hours of mechanical engineering replaced the junior year's heat power engineering courses and, the following year, six hours of civil engineering replaced half of the mechanical engineering courses.

The university returned to the two-semester academic calendar in 1925. That year's chemical engineering curriculum was as follows:

Freshman Year (All Engineering Students)	Sophomore Year (All Engineering Students)
General Physics—8 hours	General Chemistry—8 hours
English Composition—6 hours	Calculus—8 hours
College Algebra—3 hours	Mechanical Drawing—4 hours
Trigonometry—3 hours	Intro. to Mechanical Engr.—6 hours
Analytical Math—5 hours	Intro. to Civil Engineering—6 hours
Mechanical Drawing—2 hours	Intro. to Electrical Engr.—6 hours
Descriptive Geometry—2 hours	Mech. Engr. Shop—6 hours
Mech. Engr. Shop—2 hours	Military Art—2 hours
Military Art—2 hours	

Junior Year	Senior Year
Quantitative Analytical Chem.—3 hours	Physical Chemistry—6 hours
Qualitative Analytical Chem.—3 hours	Advanced Quantitative Chem.—3 hours
Organic Chemistry—8 hours	History or Industrial Chemistry—3 hours
Mechanics and Materials—10 hours	Advanced Inorganic Chemistry—3 hours
Principles of Elec. Engr.—3 hours	Thermodynamics—6 hours
Electrical Engineering Lab—1 hour	Electives—15 hours
History or Industrial Chemistry—3 hours	
Electives—5 hours	

Nine of the twenty hours of electives had to come from other chemistry courses. The next changes to the curriculum came in 1928, when Dynamo Machinery replaced Principles of Electrical Engineering in the junior year and six hours of Economics or Business Administration were added to the senior year.

Perhaps the most important change made in the course offerings came in the 1932–1933 school year. That year's catalog notes a new course, Elements of Chemical Engineering. After thirty years, chemical engineering students at the U of A could finally take a course in their own engineering field. Elements of Chemical Engineering later became that fundamental foundation for all chemical engineering students, Unit Operations. Students used the *Elements of Chemical Engineering* textbook by Badger and McCade, and Professor Harrison Hale taught the course. For some reason, the course was not listed as a requirement in the chemical engineering degree plan for that year, possibly because the course was new and, although it was listed as being offered, had not been added to the degree plan.

The 1934–1935 curriculum introduced a chemical engineering-specific sophomore year, including five hours of General Chemistry, one hour of Engineering Chemistry, three hours of both Quantitative and Qualitative Analysis, eight hours of Calculus, four hours of Mechanical Drawing, six hours of Principles of Economics, four hours each of English and Public Speaking, and two hours of Military Art. The junior- and senior-level course load remained the same as the preceding year's, with the exception of Elements of Chemical Engineering, which became a required course for chemical engineering majors.

From 1934 through 1936, the administration made no substantive changes to the chemical engineering curriculum. The arrival of Instructor of Chemistry Ralph Higbie, who had extensive practical experience in the chemical industries, however, brought new changes; the 1937–1938 course catalog noted two new courses, Chemical Engineering Laboratory, which acquainted students with "principles and practices of applied chemical engineering," and, for the first time at the U of A, a course called Unit Operations, designed to teach the basic "principles and practices of chemical engineering."

Besides the important curricular modifications, other changes, notably to the physical setting, greatly impacted chemical engineering students of the 1918–1938 period. When Professor Hale arrived at the U of A in 1918, the Chemistry Building, where most chemical engineering classes met, was already out of date; students squeezed into overcrowded classrooms and used outmoded laboratory equipment. In order for the programs in chemistry and chemical engineering to develop, a new building would be needed. Despite the obvious deficiencies of the old building, shortage of funding required the necessary work to be put off for over a decade and a half.

In 1925, however, the state legislature appropriated money for the construction of a new engineering building—Engineering Hall, as it is known today. The new engineering building, along with a new agricultural building, would be the first to be created under a new master layout plan for the university.

The three-level building, designed by the St. Louis architectural firm Jamies and Spearl, arose on the southern edge of campus, near the intersection of Dickson Street and University Drive. The basement contained laboratories, an instrument room, a repair shop and tool room, and the janitor's office. The first floor, with twenty-one rooms, contained a 209-seat auditorium, a 10,300-volume library, the offices of the dean and the department heads, and eight classrooms. Drafting rooms, a radio laboratory, two classrooms, faculty offices, a blueprint room, and an art studio made up the second story.

According to the *Arkansas Engineer,* the new Engineering Hall was "fireproof and of a construction capable of being added to in the future with very little expense. It will give the engineering college a chance to expand, which it has long needed, and with the addition of new and modern equipment will make it the most complete and efficient college on the

Engineering Hall

campus." Although the building housed the departments of civil, electrical, and mechanical engineering—chemical engineering was still under the department of chemistry—any engineering classes required for the B.S.Ch.E. degree would have met in the new building, which opened in 1927.

As an added bonus for chemical engineers, the College of Agriculture vacated Gray Hall to move into its own new building in 1927, thus allowing the chemistry department to take over part of the old building for Dr. Edgar Wertheim's organic classes and laboratories. The old Engineering Building, built in 1905, became the Commerce Building, housing the departments of philosophy and psychology, military art, public speaking, and the School of Business Administration.

By 1930, the need for a new chemistry building had become overwhelming. "The next most pressing need of the University is to construct and equip at least two buildings for the sciences," President John C. Futrall reported in 1930. "Not one of the great basic sciences of chemistry, physics, botany, zoology, and geology is by any means adequately provided for, either in housing or in laboratory equipment. The most serious condition is in the department of chemistry which is still using a building erected 25 years ago to accommodate 75 students. There are now almost ten times that many students in the department." By this time students and faculty had to perform their lab work in separate buildings, and quizzes and recitations were given in five buildings scattered around campus. Although the Great Depression's onset left little funding for new buildings, the state legislature appropriated money for one in 1931. These funds were later withdrawn, however, in the face of budget shortfalls.

When Franklin Delano Roosevelt took over the presidency in March 1933, he immediately pressured Congress to pass his economic relief and recovery efforts known collectively as the New Deal. Among the New Deal programs passed in 1933 was the Public Works Administration (PWA), designed to provide relief for unemployed persons by creating billions of dollars' worth of public buildings, highways, and other important structures. The university applied to the agency on September 18, 1933, and, in May 1934, the PWA agreed to provide loans for the construction of three university buildings: a Medical School building at Little Rock, and library and chemistry buildings at Fayetteville. After well over a decade of pointing out the deplorable condition of the old building, the chemistry department, and its chemical engineering branch, was getting a new setting.

The groundbreaking took place in July 1934. Faculty and students placed an autographed copy of the *Hexagon,* official publication of Alpha Chi Sigma, in the cornerstone. The building, designed by the Wittenberg and Deloney architectural firm of Little Rock, was designed to be integrated into the campus master plan created in 1925. As such, it was designed as a companion to the Agricultural Building, providing what was planned as a southern boundary of campus. According to an article written by Harrison Hale in the *Arkansas Engineer,* "the building is located just north of the heating plant and due south of the Agriculture Building. It is a companion to this, being the same length, 257 feet, on its north form, and the east wing is the same length as the east wing of the Agricultural Building, 105 feet. Midway between the two the new Library is being built."

The three-story (plus basement) gray limestone building, which still houses the university's Department of Chemistry—and which housed chemical engineering for a decade or so—faces north. Besides some basic maintenance work, very little has changed within the building to this day.

Chemistry Building. (*Razorback,* 1947)

Depression-era students found themselves in a lobby upon walking through the building's northern entrance. Directly across the east-west hallway bisecting the building was the auditorium, capable of holding 277 students and including a sloped floor for unobstructed views of the professor. The main floor also included a classroom, storeroom, the general chemistry laboratory, freshman balance rooms, laboratory preparations rooms, faculty and departmental offices, individual student laboratories, and faculty laboratories. The upper floors included the qualitative and organic chemistry laboratories, placed directly above the general chemistry laboratory for convenience. The qualitative analysis laboratory and an advanced organic laboratory were stacked on top of each other on the east wing's second and third floors. Each floor had storerooms connected by a freight elevator. Students could study in the second-floor's library and reading room.

The new Chemistry Building's basement became the realm of the chemical engineers. It included a lecture auditorium in the east wing of the building, "planned especially for their [chemical engineers] convenience, the east entrance being nearest to the engineering building and leading directly to this lecture room." The west wing of the basement included laboratories for industrial chemistry and chemical engineering. Not only did the chemical engineers have their own laboratories and classrooms for the first time, but they also shared the overall educational and safety benefits of moving into the new building. As Professor Hale wrote: "It is doubtful if the former student would feel entirely at home when he first worked in the new building. The balances will not be so close that they touch and the balance support will be steady; laboratories will not be overcrowded; [and] hydrogen sulfide smell and acid fumes should be absent due to fans located in a fan room in the attic."

The Chemistry Building, along with the new library, was dedicated at commencement exercises in 1935. Dr. Edward Barrow, president-elect of the American Chemical Society (ACS), was the chief speaker at the dedication ceremonies.

Chemical engineering students of the twenties and thirties participated in a variety of extracurricular activities, some new, some old. Engineers' Day continued to be a highlight of the school year for them, although many changes took place in the annual event.

The change in the academic calendar to a term system during World War I had a direct impact on Engineers' Day in that final exams for the winter term were given the same week as St. Patrick's Day. Engineers' Day was thus moved to the second Tuesday of the spring term (usually in early April). The new date was used until the university returned to the semester system in 1927. Also during World War I, the student engineers created one of the day's most appealing attractions: an electric railway that was set up in the engineering building's basement. "The railroad proved to be a very popular attraction but is said to have been rather hard on the doors in the north end of the hall at times when the motorman was slow about getting the car stopped," noted one report. In 1920, the railroad was set up to run between Old Main's south entrance and the main entrance of Engineering Hall.

Engineers' Day, 1920s. The Bucking Ford regularly appeared at Engine Day during the 1920s. On May 4, 1928, it was lost when the University shops were destoyed by fire. (*Razorback,* 1925)

The engineers also placed an electric "Engineering" sign on Old Main's south tower and provided a fireworks display. More notable, at some point during these early years of the celebration, a band of merrymaking engineers took to painting shamrocks (in honor of Saint Patrick, the "patron saint of engineers") in conspicuous places about campus. Perhaps the most gaudy shamrocks found their way onto the Agriculture Building, a reflection of a growing and enduring competition between the agris and the engineers. In what became a twice yearly occurrence, once on Agri Day and once on Engineers' Day, the celebrants tried to deface their enemies' building while defenders valiantly tried to protect it.

Occasionally, the participants took things too far, as in 1935, when an agri-engineer rock fight broke out. As the *Arkansas Engineer* noted, "Someone threw a rock. Who it was will never be known for some say that an Engineer started it and others are just as positive that it was the boys in the other building. Chances are that it was someone that belonged to neither college. But a rock was thrown; and not only one rock but enough to break more than a dozen windows. Every one in the fight had a good time, including the Engineer that had a brick dropped on his head[,] and everyone went home happy." Although they went home happy, the participants in the melee could not have been pleased the following morning, when the university's business office announced that the Engineers' Day dance would be cancelled unless all damages were paid for. Committees from both colleges met and agreed to temper their celebrations in the years to come.

The chemical engineering students did not relegate their extracurricular activities only to Engineers' Day, however. Other outside opportunities for social and instructional activities beckoned as well, the most notable of these being Gamma Chi (later, Alpha Chi Sigma) and the Arkansas Institute of Chemical Engineers (AIChE).

In 1905, Professor Charles G. Carroll created the earliest precursor to a local chapter of Alpha Chi Sigma, the national chemistry fraternity, on the U of A campus when he founded the Journal Club in 1905. In 1917, at Professor James S. Guy's urging, the Journal Club restructured itself into the Chemical Club, a more formal organization which held biweekly meetings "to discuss some current topic of interest." A year later, in early 1918, the Chemical Club created a constitution and bylaws and changed its name to Gamma Chi; the organization's growth was stunted, however, by the fact that almost all members were serving in the armed forces by the fall of 1918.

Following the armistice of November 1918 and the return of most of the chemistry students to campus, Gamma Chi revived. The year 1919 witnessed a flurry of activity on the group's part. Following the prewar organization's example, the group held biweekly meetings and added a social component—a premeeting dinner—to its activities. Gamma Chi also adopted a ritual and created a formal process for choosing members, which allowed the group to receive official recognition from the university.

Perhaps most important, in February 1919, Gamma Chi sent "literature, information[,] and letters of recommendation" to the Supreme Council of Alpha Chi Sigma for consideration as a possible local chapter of the national organization. More than a year and a half later, in November 1920, Alpha Chi Sigma sent word that an official inspection would occur. On December 15, 1920, a representative of Alpha Chi Sigma, C. W. Seibel, met with the members of Gamma Chi. Three months later, in March 1921, the Supreme Council granted Gamma Chi the right to submit a formal petition asking for a charter and the installation of an Alpha Chi Sigma chapter at the University of Arkansas.

Gamma Chi, 1919. (*Razorback,* 1919, 175)

Gamma Chi prepared and sent its petition in the spring of 1921. The petition began with a formal greeting:

> We, the undersigned members of Gamma Chi, being members of the faculty and students of good standing in the Department of Chemistry, believing that the presence of such an organization as Alpha Chi Sigma would be of benefit not only to ourselves individually but also to the

> Chemical Department of the University, do hereby petition the granting of a charter and the installation of a chapter of Alpha Chi Sigma at the University of Arkansas.

Faculty petitioners included Harrison Hale, head of the chemistry department; John William (J. W.) Read, head of the agricultural chemistry department; and Harvey McCormick Trimble and Allen Sparrow Humphreys, instructors in the chemistry department and members of Alpha Chi Sigma. Student petitioners were George F. Blodgett, James C. Colbert, Sterling B. Hendricks (chemical engineering), Felix A. Kimbrough, Bert H. Lincoln, Ralph E. Maxwell, Calvin H. McDaniel (chemical engineering), R. Edwin O'Kelly, Bryan B. Paul, Nat L. Shepard (chemical engineering), John F. Smith, L. J. Williams (chemical engineering), F. S. Woodward, and Elmer G. Wakefield.

The petition also included a one-page description of the University of Arkansas. Detailed descriptions of the departments of chemistry and agricultural chemistry included biographical sketches and photos of the faculty and student members of Gamma Chi, attendance records, and a list of courses offered. Finally, recommendations in support of Gamma Chi's petition were attached.

Despite Gamma Chi's best efforts, the Supreme Council did not grant Gamma Chi a charter. On May 8, 1922, Ben H. Ball, secretary of Alpha Chi Sigma, notified Gamma Chi that "the sentiment among the chapters seems to be that Gamma Chi petitioned a little too soon." He added that "it would be difficult to support a chapter with the proper numerical strength at the University of Arkansas."

For the next five years, Gamma Chi continued to grow, as did the university and the chemistry department. Such growth did not go unnoticed; an April 4, 1927, letter noted that "Dr. W. S. Ritchie would find it convenient to make Gamma Chi an official visit if the organization was still interested in Alpha Chi Sigma." That visit occurred on May 7, 1927. On June 8 of that year, Gamma Chi received word that it could once again petition the Supreme Council.

The second petition included the same formal greeting as the first. Faculty petitioners included Hale, Read, Humphreys, Lyman E. Porter, and Edgar Wertheim. Student petitioners were Jacob R. Meadows, Earl Hays, Phillip Schmitt, Harold Steele (chemical engineering), William E. Gann, Charles E. Caldwell, Hugh B. Estes, Robert Kimbrell, Hugh M.

Boggs, Fontaine Earle (chemical engineering), Cleveland Hollabaugh, Lyle Alexander, Louis T. Byars, Gaston F. Bell, and Jim McKenzie. In addition to information on and photos of Gamma Chi members, the chemistry department, and the university, the petition included a brief description of Gamma Chi alumni and letters of recommendation.

Upon their return from the Christmas 1927 holidays, members of Gamma Chi finally received the news they had hoped for: Alpha Chi Sigma had granted them a charter. The official installation took place on April 21, 1928, and was presided over by Walter S. Ritchie and Robert Boucher, both of the University of Missouri. On a day described as "exceedingly rainy," Ritchie, Boucher, and Gamma Chi alumni and current members gathered in front of the Chemistry Building for photos, then moved to the new agricultural building for the actual installation ceremony, which was conducted by Ritchie, Boucher, Porter, and Wertheim.

Following the installation of the Alpha Sigma chapter, C. D. Caldwell, Arthur Hale, Robert Kimbrell, Wycliffe Owen, and John Womack were elected as officers. That evening, the new members of Alpha Chi Sigma enjoyed a dinner prepared by the university's home economics department while listening to speeches on the histories of Gamma Chi and Alpha Chi Sigma. A report to *The Hexagon of Alpha Chi Sigma* noted that "Alpha Sigma faces a splendid future in the University of Arkansas, and it is firmly

Alpha Sigma Chapter of Alpha Chi Sigma, 1929. (*Razorback,* 1929, 281)

resolved to make this chapter one of the best in Alpha Chi Sigma. . . . We are the baby chapter now but we intend to remain so only in point of age."

By the early 1930s, chemical engineering students at the University of Arkansas recognized the need for an organization dedicated to their particular field of expertise. Thus, in 1931 the students created the Arkansas Institute of Chemical Engineers, modeled on the national American Institute of Chemical Engineers. According to the 1934 *Razorback,* the AIChE "is a an organization for students of Chemical Engineering who are particularly interested in topics of everyday chemistry and in the advancement of that science. Meetings of A.I.Ch.E. feature papers and discussion on subjects of general interest to chemists and also give opportunity for carrying on interesting chemical experiments." Chemical engineering students who attended at least three-fourths of the meetings during their junior and senior years and presented one or more original papers received credit for Chemistry 501, "Chemical Engineering Seminar." As the decade wore on, and the number of chemical engineering students increased, the size of the group grew as well.

Outside of their particular chemistry/engineering domain, chemical engineering students in the years 1918 to 1938 continued to be active participants in university life. All students entering the University of Arkansas in 1918 found themselves in a situation very different from students of just a few years earlier. The Great War wrought a number of changes on the life of the university student. Total enrollment for the university in 1918 included 1,017 students plus 300 soldiers receiving vocational training. Students came mostly by railroad from sixty-nine of Arkansas's seventy-five counties and from twelve states. Costs, including room, board, laundry, and books, ranged from $318 for the most frugal students to $478 for those with expensive tastes. Forty-five students received degrees at the end of the 1918–1919 school year, a decrease of one from the year before and twenty-one from two years before.

During the war, the university witnessed a drain of service-age men from the campus. By Armistice Day, some 1,022 student had entered the military, as had 35 faculty members. Of those, 29 died. Additionally, "several hundred members" of the Student Army Training Corps (SATC) were scheduled to enter active duty when the war ended. Students who did not enlist participated in any number of other war-related activities, including buying Liberty Bonds and War Savings Stamps.

Student Chapter—The American Institute of Chemical Engineers, 1936. *l. to r., starting at top row:* Loyal R. Babb, James F. Bourland, Robert W. Brown, Glen Burleson; Paul F. Bustion, Gerald W. Chastain, Meyer M. Cooperman, William T. Cravens; Franklin K. Deaver, Coy A. Hardcastle, Alexander E. Harris, Lynn D. Howell; Ashley C. Johnson, Maynard P. Johnson, James P. Lea, Edgar G. Lowrance; Allen C. Mark, Warren O. Nance, William L. Nelson, J. F. Norman; John L. Oswalt, Howard E. Powell, Ralph D. Skinner, Winston W. Treece; Scott W. Walker, Claud S. Wilson, Julius J. Woodruff. (*Razorback,* 1936, 223)

Just about all male students became members of the Student Army Training Corps, created in the fall term of 1918. "Any young man who was registered for the draft on September 12, 1918, who was physically qualified and who had completed the necessary high school work, was eligible to enter," noted University President John C. Futrall in a December 1918 report. The creation of the corps resulted in the university becoming a quasi-military installation, as Futrall noted: "The contract with the Government provided that the University should feed the men and house them in barracks so that they might be kept under military discipline at all times, and should give them instruction in subjects regarded as necessary for the training of the men." Guards patrolled the university's perimeter; no person could enter campus without a pass. The student-soldiers received complete uniforms plus a stipend of thirty dollars per month. The federal government paid for the construction of five barracks to house the soldiers, plus two mess halls, bathhouses, and a few other small buildings for SATC purposes. The government also mandated that the school year be divided into three (fall, winter, and spring) terms of three months each.

Campus Life during World War I. (Simpson, 80)

The end of the war, which came on November 11, 1918, brought about a rapid end to the SATC. The U.S. government declared that all members of the SATC were to be discharged by December 21, 1918. The brief stint of the SATC at the University of Arkansas had shown only a glimpse of what could have been. "If the war had continued the experiment would probably have accomplished its purpose, so far at least as the fitting of men to enter Officers' Training Camps was concerned," noted Futrall. The university's president had doubts about the SATC's benefits in the short term, however. "From an educational standpoint the experiment has been entirely unsatisfactory," he wrote. "It has proved conclusively that the routine of strict and uninterrupted military discipline, to which S. A. T. C. students were subjected, is entirely inconsistent with high grade college or university work. It was the understanding that the men should devote eleven hours per week to military work and forty-two hours per week to academic work. . . . In actual practice, however, these figures were almost reversed."

The fall of 1918 also brought an influenza epidemic to the university. The outbreak of the Spanish flu killed 22 million people worldwide, double the number killed in the Great War. It struck campus on October 9, 1918, immediately after the opening day of classes. Over the next few days 337 members of the SATC fell ill, as did 54 young women in Carnall Hall and "a considerable number of civilian students living or boarding in the city of Fayetteville." Classes were delayed while physicians and volunteers tended to the sick. Twelve members of SATC and one civilian student died before the epidemic subsided in early November.

With the Great War's end, campus life returned to the pattern of the prewar years. Although the three-term school year remained in effect, the SATC rapidly dissolved, and students began to once again concern themselves with their studies and the social activities of university life.

Some students lived off campus in private residences around Fayetteville. Women still lived in Ella Carnall Hall. The men continued to live in Buchanan Hall, Hill Hall, and Gray Hall. The dormitories were the sight of many pranks and exciting antics, remembered one student who attended the university in the 1920s: "Horses used to graze on the campus, and my roommate and I caught one of the horses. . . . a guy in the dorm was at a dance, and we knew he was. We put that horse in his room. And he had a narrow room, and the back end of the horse was right in over his bed."

The literary societies that had been the center of most student extracurricular activity during the university's first five decades rapidly died out as a result of two developments. First, coursework assigned by faculty came to allow more freedom of choice for students, so that they no longer felt the need for outside sources of intellectually stimulating material. Second, other organizations and activities, such as Gamma Chi and the Arkansas Institute for Chemical Engineers for chemical engineering students, became much more central to student life by the 1920s. Additionally, as students came to be accepted as "sophisticated members of the general community," they increasingly had "access to the normal pleasures and entertainments of the community."

In June 1922, the University of Arkansas celebrated its semicentennial. Two thousand out-of-towners descended upon Fayetteville (the railroad companies offered special reduced rates from most Arkansas towns) for five days to commemorate "Fifty Years of Service," as the celebration's motto said. Highlights included a pageant, written by university English professor Jobelle Holcombe, "portraying the history of the University and its service to the state," the lighting of "the Great White Way," a new system of streetlights connecting campus and the town square, and addresses by distinguished alumni, including M. L. Bell (class of 1898), vice president of the Rock Island Railroad, and U.S. Senator Joseph T. Robinson (class of 1891).

The military continued to play a large role in the lives of university students after World War I. The Reserve Officers' Training Corps (ROTC) had been created in 1916; Scabbard and Blade, a military honor society, became one of the top campus organizations; and the military ball became an annual highlight of students' social calendars. Upper-level students in ROTC classes received a small monthly stipend in addition to being eligible for reserve commissions as second lieutenant upon graduation. The stipends, always attractive, became objects of stiff competition after 1929, when the depression that gripped the nation made any income significant.

The YMCA and YWCA continued to be important social and spiritual venues for students in the twenties and thirties. The construction of a YMCA "Hut" as part of the Student Army Training Corps construction projects allowed the two organizations, as well as most other student activities, a central location for offices and activities. Perhaps the most significant addition to the YMCA in the aftermath of the Great War was the new secretary, William S. Gregson. Gregson remained at the university for more than a quarter century, serving as university chaplain and YMCA secretary until 1946. His guidance left an indelible stamp on the lives of thousands of students.

Athletics also continued to play an important role in the lives of university students. Following Hugo Bezdek's resignation as football coach and athletic director in 1912, the Razorback football teams had posted mediocre marks. The team reached perhaps its lowest point in 1918, however, when Coach Norman Paine's team, depleted by the loss of most of its members to the SATC, lost at Oklahoma by a score of 103–0. Two more coaching changes marked the next four years. Finally, in 1922, Coach Francis Schmidt began a seven-year tenure that brought respectability back to the program. Schmidt left to become football coach at Texas Christian University (TCU) at the end of the 1928–1929 school year. His replacement, Fred Thomsen, served thirteen years and became one of the most popular coaches in Arkansas history. Highlights of Thomsen's years include the first All-American Razorback, Wear Schoonover; Arkansas's first victory over the University of Texas (UT) and first appearance in a bowl game in 1933; and the Razorback's first Southwestern Conference championship in 1936.

Coach Schmidt put together the first men's basketball team in 1923. After two years of modest success, the team came together and won the Southwestern Conference championship five years in a row. The wins tapered off, however, when Schmidt left for TCU. In 1933, Glen Rose arrived on campus as basketball coach and revived the team's winning ways, winning three conference championships over the next five years.

Arkansas's baseball teams had only modest success during the 1920s, posting records that rarely went above .500. When Schmidt left for TCU in 1929, the university dropped baseball as an intercollegiate sport due to poor attendance and financial losses from traveling expenses (the closest conference rival was three hundred miles away). The University of Arkansas would have no baseball team until 1947.

Facilities for athletic events continued to be improved between 1918 and 1938. In 1923, a fifty-by-one-hundred-foot frame building became the first field house for the basketball team. Known as "Schmidt's Barn," it was located in what is today the open area between Mullins Library and the Arkansas Union, but at the time it was in the center of the university's athletic facilities, including the football and baseball field, the track, and the tennis courts. A Girls' Gymnasium (today's Army ROTC building) was completed in 1924. The football field, which ran east and west, was turned north-south in 1926. In 1936, a new Men's Gymnasium (today's University Museum) replaced Schmidt's Barn. Two years later Bailey Stadium for football (today's Reynolds Razorback Stadium) was completed in a valley to the west of campus.

The financial depression that began in October 1929 affected students at the University of Arkansas in a number of ways. University income decreased dramatically, and, as it did, faculty salaries dropped, precipitating a faculty exodus in search of better-paying jobs. Although enrollment wavered for the first few years of the depression, the number of students on campus increased steadily from 1933 to 1938, probably a reflection of the inability to find jobs.

The most noticeable changes came in the form of new physical structures on campus. As part of the relief aspect of his New Deal, President Franklin Roosevelt had sponsored the creation of the Public Works Administration, which offered discount loans for the construction of public projects such as schools, courthouses, and post offices. The PWA sponsored the construction of nine University of Arkansas buildings: the Chemistry Building, the University Library, and a new Medical Sciences building; the Men's Gymnasium, Razorback Hall (now Gibson Hall), Razorback Stadium, a Student Union (Memorial Hall), a classroom building (Ozark Hall), and the Home Economics Building.

University students during the twenties and thirties could engage in a wide array of off-campus activities, and thanks to the city's paving efforts (Dickson Street received its first pavement in 1918; the square in 1924; Highway 71 in the 1930s), getting off campus became even easier. Attending the movies on the square at either the Palace Theater or the Royal Theater became a top choice for entertainment, and swimming at Trent's Pond always provided an enjoyable way to pass the time on a hot afternoon. The Majestic Café (George's) became a favorite of university faculty and students during the 1930s; almost any university student could run a tab, and faculty members always knew that Table One was reserved for them. Sports fans who needed even more excitement than the Razorbacks could provide ventured out to support the Fayetteville minor-league baseball team.

Fayetteville, 1920s-1930s. (Courtesy Peter Harkins / Shiloh Museum of Ozark History)

Fayetteville itself continued to grow during this period. In 1928, the town celebrated its centennial with a pageant parade at Wilson Park exhibiting various milestones in its history. That same year, Drake Field began operations as the local air strip; KUOA, the city's first radio station,

Dickson Street, 1933. (Courtesy Joe Neal / Shiloh Museum of Ozark History)

began operation on the U of A campus a year earlier. Hotels, which had always existed on or near the square, began to pop up along Dickson Street—among them the OK Hotel, the Shady Lane, the Midway, and the Winkleman. The financial depression of the thirties hurt the local economy, but not to the extent it did other towns; there were no soup lines, no banks went belly up. Business slowed, but it did not stop. Besides the new buildings on campus, the Works Projects Administration (WPA) built Fayetteville a drainage system and increased the amount of concrete paving on the town's streets. The Civilian Conservation Corps (CCC) built trails and campgrounds for outdoor enthusiasts at what is today's Devil's Den State Park, just an hour south of Fayetteville.

Although Professor Harrison Hale continued as chair of the Department of Chemistry, the year 1938 marked a turning point for the U of A chemical engineering program. That year a young professor arrived to become the university's first instructor with a degree in chemical engineering. His name was Stuart McClain and his brief tenure at the university marked the beginning of a new, transitional period for the program.

Biographical Sketches, 1918–1938

Harrison Hale replaced James S. Guy as professor and chair of the Department of Chemistry in 1918. Hale would be the dominant figure in both chemistry and chemical engineering at the University of Arkansas for the next two decades. Born in Columbus, Mississippi, Hale received his bachelor's of arts from Emory College, his master's of science from the University of Chicago, and his Ph. D. from the University of Pennsylvania, where he was a two-year Harrison Fellow in Chemistry. In 1936, he added a doctorate of law from Drury College. He was a member of Alpha Chi Sigma and Phi Eta. Hale served as professor of chemistry until 1945, when he was named emeritus professor. He also served as a member of the Bureau of Research staff.

Professor Hale taught a wide variety of chemistry classes at the

Chemistry Faculty, 1940. *l. to r., front row,* Allen S. Humphreys, Harrison Hale, unidentified, Lyman E. Porter; *back row,* Edgar Wertheim, Warren H. Steinbach, Andrew D. Bogard, Ralph W. Higbie. (Mullins Library Special Collections, photo collection #167)

U of A, reflecting his commitment to both technical and popular aspects of his field. Over the years, Hale's course offerings included General Chemistry, Industrial Chemistry, History of Chemistry, Advanced Inorganic Chemistry, American Chemistry, Food Analysis, Special Methods in Quantitative Analysis, Chemistry Seminar, General Chemistry for Engineers, Chemical Research, Elements of Chemical Engineering, Everyday Chemistry, Descriptive Chemistry, and Survey of Science.

Hale had the rare ability to inspire both as a person and as a professor, as one of his students remembered:

> When I graduated from high school, I was inclined toward studying medicine, so I went to the university. Of course, I could go there cheap, on in-state tuition. . . . I'll never forget how I got interested in chemistry and stayed with it. . . . I had gone in that first day I was there and gotten my books, an armload. And I was coming out of Charlie Stone's bookstore, down those stairs, and I saw this gentleman, walking along,

> as I walked by Old Main. And, as I approached him, he smiled very genially and he said "that's a big load of books you got there young man. I hope you can manage it." And I walked on and I think in my mind who he might be, and I thought, "yes, he might be a professor." So I took my books on to Buck [Buchanan] Hall. . . .
>
> The next Monday morning . . . I had freshman chemistry. And there behind the lectern, in the chemistry building that's still up there, stood the man that I met on the sidewalk . . . Dr. Harrison Hale. And I fell in love with him and with his chemistry.

Before arriving at Arkansas, Hale had been city chemist for Springfield, Missouri, consulting chemist for Fort Smith Municipal Water Works, and a chemist with the Missouri Food Commission. He also served as president of the Fayetteville Chamber of Commerce. His university men's Sunday School class of the Central Presbyterian Church became famous for their attendance marks, regularly beating classes of other denominations and other areas in good-natured attendance contests. In a statement that sums up the feelings most had for Hale, a former student stated, "Dr. Hale was one of the kindest, gentlest people I've ever known."

Edgar Wertheim became associate professor of chemistry at the University of Arkansas in 1921. In 1929, he was named professor of chemistry. Wertheim had received a bachelor's of science degree from Northwestern University, a bachelor's of physical education from the YMCA College of Chicago, a master's of science from Kansas University, and a doctorate from the University of Chicago. Upon Harrison Hale's retirement in 1945, Wertheim became head of the chemistry department, a post he held six years. In 1951, he was named emeritus professor.

For more than two decades, Wertheim instilled fear into hundreds of organic chemistry students. "Lord, he was worse than the devil himself," remembered one former student. "He could terrify his classes. He always lectured with his back to the class . . . writing all these complex organic formulae on the board, and he would draw with one hand and come across with the eraser in the other and wipe them off." The benefit of Wertheim's demanding teaching style was that students came out of his classroom with a firm knowledge of organic chemistry. "We hated Dr. Wertheim when we were in his class," noted another of his students. "We thought he was an arrogant ass—and he was to some extent. It was only after I got out of college and was dealing with people from [places like] Georgia Tech, University of Texas . . . that [I realized that] I knew twice as much organic chemistry as any of those guys. I really did. And I finally got through with a C. So I look back and say I didn't like him at the time, but he was one of the most effective professors I ever had." Another student remembered that "he was a great motivator to we fledgling chemical engineers, but he was very tough."

Wertheim used his own well-regarded book, *Textbook of Organic Chemistry*, for his organic classes; a former student noted that "one needed a wheelbarrow to tote his organic chemistry book and lab manual around." His classes included Elementary Organic Chemistry, Organic Chemistry, Advanced Organic Chemistry, Organic Quantitative Analysis, Organic Preparations, Special Methods in Quantitative Analysis, Special Organic Chemistry, Chemical Research, Organic-Physiological Chemistry, Graduate Chemical Research, and Advanced Organic Synthesis. He was a member of *Deutschechemische Gesellschaft* and was faculty advisor for Alpha Chi Sigma.

Allan "Squire" Humphreys became instructor in chemistry at the University of Arkansas in 1918. He arrived on campus immediately following his discharge from service during World War I. Humphreys had received a bachelor's of science from Drury College, a master's of science from the University of Pennsylvania, and a bachelor's of pedagogy from Southwest Missouri State Teachers College. He was named assistant professor of chemistry in 1921, and, in 1937, Humphreys became personnel director for men, a position similar in rank and duty to dean of men. He attained the ranks of associate professor in 1941, full professor in 1944, and emeritus professor in 1955.

Humphreys was a member of Alpha Chi Sigma, Phi Lambda Upsilon, Gamma Alpha, and the American Chemical Society. In the classroom, he instructed students in classes such as Elementary Organic Chemistry, Organic Chemistry, General Chemistry, Inorganic Preparations, General Chemistry for Engineers, Chemical Calculations, and Oxidation-Reduction Reactions.

"Squire Humphreys was one of the best teachers I ever had," recalled one of his students. "To me, he was a great teacher," remembered another, "but beyond that he was a personality that could command respect and yet come down to [our] level. . . . Everybody loved him."

Lyman Porter became instructor in chemistry at the University of Arkansas in 1921. He rose to assistant professor in 1927 and associate professor in 1937. Porter received his bachelor's, master's, and doctoral degrees from Yale University. Prior to his arrival at Arkansas, Porter held a position with Barrett Roofing Company. He was a member of Alpha Chi Sigma.

At one time or another, Porter taught Qualitative Analysis, Advanced Qualitative Analysis, Quantitative Analysis, Advanced Quantitative Analysis, Physical Chemistry, Electro-Chemistry, Metallurgy, Special Physical Chemistry, and Special Methods in Quantitative Analysis.

Students remembered Porter as a kindly person, both inside and outside the classroom. "I didn't take any of his classes, but I worked for him one semester, fixing various mixtures that he needed in his classes," remembered one chemical engineer. "He was crippled, I mean he really was badly crippled, and I don't know if he had had polio or been in a wreck, but he was badly, badly crippled. But, pleasant. I mean, he had a smile on his face all the time, and I enjoyed working for him. He was an easy guy to work for."

Ralph W. Higbie became the first true chemical engineer on the faculty when he came to the University of Arkansas in 1936. He received his bachelor of science, master of science, and doctor of science degrees from the University of Michigan. Previous to his appointment as instructor in chemistry, Higbie had worked at the research laboratory of the Eagle-Picher Lead Company of Joplin, Missouri. His arrival at the U of A provided a "boon" to the chemical engineering students and opened the door for the creation of a local chapter of the American Institute of Chemical Engineers.

One former student, Leonard W. Russum (B.S.Ch.E., 1939), remembered Higbie as "an intense, very curious, jumpy individual, [who] accompanied the junior engineers in-or-near 1937, on their spring tour of Arkansas industry." Russum also recalled that Higbie was quite the city boy and was unprepared for some of the rigors of Arkansas life: "While in

Crossett at the lab of the Crossett Chemical Company, Dr. Higbie, a city-bred fellow, had to visit the outhouse. While inspecting the system, his Parker pen fell in." Russum added, "he called for help in fishing it out, to the merriment of all."

Chapter Three

Transitions, 1938–1948

The Great Depression still wracked the United States at the beginning of the 1939–1940 school year. The nation, however, faced a new problem: what to do about increasingly complex and foreboding rivalries in Europe and Asia. Militarists in Japan had seized control of the government and begun preparations for Asian conquest in 1936. That same year, Italy, under the fascist regime of Benito Mussolini, invaded Ethiopia in a bid to enlarge its control of northern Africa, and Germany, under the leadership of National Socialist Adolf Hitler, reoccupied the Rhineland demilitarized zone that had been set up between France and Germany following the Great War.

On September 1, 1939, Hitler's forces invaded Poland, and a second World War began. The Axis powers—most notably Germany, Italy, and, later, Japan—squared off against the Allies—including Britain, France, and, later, the Soviet Union. In the United States, debate raged over the nation's neutrality.

By 1941, the United States was actively providing economic aid to the Allies in Europe, and U.S.-Japanese relations had reached a point of no return. When the attack on Pearl Harbor came on December 7, 1941, the only real surprise was its location. From this point on, the United States became an active belligerent in World War II.

World War II had a profound impact on the United States. More than 16 million Americans were mobilized and casualties amounted to more than 1 million—407,000 dead; 672,000 wounded. Defense expenditures reached $560 billion. At home, the war did what President Franklin Roosevelt's New Deal could not—it cured the Great Depression. The war also opened up new opportunities for women to join the work force and for minorities to make gains in their social status.

The war dramatically affected higher education in the United States, too. The wartime emphasis on scientific and technological developments carried over into the postwar period, when a huge influx of new students brought many universities to the bursting point. Using the GI Bill or

Public Law Sixteen for funding, returning veterans enrolled in colleges; a university education came to be seen less as a luxury and more as a necessity for life in the modern world. At the University of Arkansas, overall enrollment reached 5,566 by the 1947–1948 school year, up from 2,618 in 1939–1940. Chemical engineering graduates rose from thirteen in 1939 to a peak of fifteen in 1943, then to lower levels throughout the duration of the war. By 1947, the number of graduates had regained its prewar levels.

The 1939–1940 school year marked the earliest time the course catalog listed the chemical engineering curriculum separate from chemistry; that year chemical engineering was listed under the College of Engineering. Six years later, in 1945, the separation from the Department of Chemistry became complete, and the Department of Chemical Engineering as an independent unit was born.

Within the College of Engineering, chemical engineering became, proportionally, a much more popular field than it had ever been before. Part of the reason for this increasing interest lay in the reliance of industry upon chemical engineers, as U of A president J. William Fulbright noted in 1940: "The directive science back of many of our industries is Chemical Engineering. This is true of our largest industries even though they are not strictly chemical in nature. A few of these are the refining of petroleum, paper manufacture, ceramics, the processing of vegetable oils, wood distillation, ore concentration and metal refining." Also, Arkansas's untapped potential could make the state a major player in the chemical industries: "The metallurgical industries in the Southwest are being shared by Arkansas to an increasing degree. It is well known that Arkansas possesses millions of tons of cinnabar, manganese[,] and antimony ores any one of which if properly developed by our own citizens would add great wealth to our state."

Despite these potentials, chemical engineering at the University of Arkansas also faced new obstacles, especially the lack of sufficient equipment. "One-third of our Engineering student body are chemical engineers," Fulbright added, "but we do not have the necessary equipment to train these men." These potentials and concerns would linger on the minds of chemical engineers at the university throughout the war years. Like so much else on campus between 1939 and 1947, the chemical engineering department existed in a state of flux, however, and the period is best viewed as an era of transition for the program. The faculty of the departments of chemistry, and, after 1945, chemical engineering reflected these changes.

The following list includes all chemistry and chemical engineering instructors who arrived at the U of A between 1939 and 1948. The large number of departures and arrivals points to the instability of higher education faculty not only at the U of A but also around the nation.

1. Stuart McLain. Instructor in Chemistry (1938–1940); Assistant Professor (1940–1944); Associate Professor (1944–1945). B.S.Ch.E., M.S., Ph.D., University of Michigan.

2. Charles Allen Walker. Instructor in Chemistry (1941–1942). B.S.Ch.E., M.S.Ch.E., University of Texas.

3. Odie Sylvester Jenkins. Instructor in Chemical Engineering (1942–1944). B.S., Texas College of Arts and Industries; B.S.Ch.E., M.S.Ch.E., University of Texas.

4. Peter Newport Bragg Jr. Instructor in Chemistry (1943–1944). B.S.Ch.E., University of Arkansas.

5. Sam William Thompson. Instructor in Chemistry (1943–1944). B.S.Ch.E., University of Arkansas.

6. William Long Belvin. Research Professor of Chemical Engineering (1944–1948). B.S.Ch.E., M.S.Ch.E., North Carolina State College.

7. Lawrence Hein. Assistant Professor of Chemical Engineering (1945–1947). B.S., M.S., Oklahoma A. & M. College; Ph.D., Michigan State University.

8. Joseph Walter Ranftl. Instructor in Chemical Engineering (1946–1947). B.S.Ch.E., University of Wisconsin.

9. Robert William Rowden. Instructor in Chemistry (1946–1948). B.S.Ch.E., M.S., University of Arkansas.

10. Wickliffe Skinner Jr. Instructor in Chemical Engineering (1947–1948). B.S.Ch.E., University of Oklahoma.

11. Henry Tibbels Ward. Professor of Chemical Engineering and Head of

the Department (1947–1948). B.S.Ch.E., University of Michigan; M.S., University of Wyoming; Ph.D., University of Michigan.

12. Freeland Elmer Romans. Instructor in Chemistry (1943–1944). B.S., M.S., University of Arkansas.

13. Lawrence Alvin Frye. Instructor in Chemistry (1945–1948). B.E., State Teachers College, River Falls, Wisconsin; M.S., State University of Iowa.

14. Alfred Ingle. Instructor in Chemistry (1946–1947). A.B., William Jewell College.

15. Robert George Mers. Assistant Professor of Chemistry (1945–1946). B.A., Austin College; M.A., Ph.D., University of Texas.

16. Bahngrell W. Brown. Instructor in Chemistry (1946–1947). B.A., University of Omaha.

17. Kenneth E. Conn. Assistant Professor of Chemistry (1946–1947). B.A., University of Nebraska; M.A., Indiana University.

18. Dillon O'Neal Darby. Instructor in Chemistry (1947–1948). B.A., University of Houston.

19. William K. Noyce. Associate Professor of Chemistry (1946–1951); Professor of Chemistry (1951–1966). A.B., Doane College; M.S., Ph.D., University of Nebraska.

20. Russell Arthur Schroeder. Instructor in Chemistry (1946–1952). A.B., DePauw University; M.S., Purdue University.

21. Edward Graydon Snyder. Instructor in Chemistry (1946–1947). B.S.E., University of Michigan.

22. Edward S. Amis. Professor of Chemistry (1947–1974); Professor Emeritus (1974–1988). B.S., M.S., University of Kentucky; Ph.D., Columbia University.

23. Floyd Telford Doane. Instructor in Chemistry (1947–1948). A.B., Nebraska State Teachers College; M.S., Louisiana State University.

24. William Kenneth Easley. Instructor in Chemistry (1947–1948). B.S., Carson-Newman College; M.S., University of Richmond.

The state of flux that existed in the faculty also existed in the department's curriculum. Although war had broken out in Europe by the beginning of the 1939–1940 school year, the United States had not yet become embroiled in the conflict. Thus, the drastic curricular changes that World War I brought about had not been enacted.

Throughout the transition period, freshmen chemical engineers continued to take a common course load with all other first-year engineering students. The classes they took included General Chemistry, English Composition, College Algebra, Trigonometry, Analytical Mathematics, Engineering Physics/Mechanics, Mechanical Drawing, Descriptive Geometry, Mechanical Engineering Shop, Military Art, and a General Engineering Freshman Orientation class. Sophomores, juniors, and seniors enrolled in the fall of 1939 could expect to pursue the following degree plan:

Sophomore Year

Physics 113, b, c, and 121, a, b, General Physics—8 hours
Chemistry 214, Qualitative Analysis—4 hours
Chemistry 254, Quantitative Analysis—4 hours
Mathematics 214, a, b, Calculus—8 hours
Drawing 212, 222, Mechanical Drawing—4 hours
Economic 203 a, b, Principles of Economics—6 hours
Military Art 201 a, b—2 hours

Junior Year

Chemistry 314, a, b, Organic Chemistry—8 hours
Ch.E. 313, 323, Chemical Engineering Operations—6 hours
Chemistry 404 a-b, Physical Chemistry—8 hours
M.E. 305, Mechanics; C.E. 315, Materials—10 hours
English 232; Public Speaking 202—4 hours

Senior Year

Chemistry 343, Industrial Chemistry—3 hours

Chemistry 363, Advanced Inorganic Chemistry—3 hours
E.E. 323, 333, Electrical Engineering—6 hours
M.E. 343, Ch.E. 463, Thermodynamics—6 hours
C.E. 311, 312 Surveying—3 hours
Ch.E. 341, Chemical Engineering Seminar—1 hour
Electives—14 hours (to be approved by the Dean of Engineering and the Head of the Chemistry Department)

This curriculum modified the coursework for chemical engineers. The most important additions were the senior-year elective courses, which included Chemical Engineering Laboratory, Chemical Calculations, Organic Processes, and Plant Design.

The following school year, 1940–1941, more chemical engineering classes were added: Chemical Engineering Seminar was moved to the junior year, and three hours of Chemical Engineering Calculations plus two hours of Physical Chemistry Laboratory were added to the senior year. The first mention of graduate-level classes came in the 1940–1941 catalog, when Chemical Engineering 406, Research, and Chemical Engineering 506, Research, offered research problems for "graduate students considered capable of attacking them." The year after that, seniors had to take four hours of Chemical Engineering Laboratory.

The 1942–1943 school year brought even more changes to the curriculum. Due to the war effort, the university switched to a three-term system similar to that used during World War I. The forty-eight-week term system allowed students to complete in three years studies that had taken four before the changes. Sophomore chemical engineering students now took Engineering Physics instead of General Physics, and Theoretical Quantitative Analysis became a part of their coursework. Juniors added Civil Engineering 302 and 305, Strength of Materials, to their studies. Senior classes remained the same. These changes remained in effect until the end of the war.

By the beginning of the 1945–1946 school year, the semester system had been reinstated, as had been the General Physics requirement for chemical engineering sophomores. Additionally, the sophomore-level course Theoretical Quantitative Analysis had been dropped and Chemical Engineering 262, Materials of Construction, added. Physical Chemistry Laboratory was added to the junior year. The senior-year courses included Chemical Engineering 472, Plant Design, in place of Industrial Chemistry, and five hours of electives (three of these had to be from chemical engineering courses) in place of Advanced Inorganic Chemistry, Physical Chemistry Laboratory, and Surveying.

Although no changes were made to the curriculum the following year (1946–1947), the Engineering Catalog carried a new descriptive note on the purpose and design of the chemical engineering curriculum. "The courses in Chemical Engineering are designed to give the student a knowledge of the fundamental engineering principles which are common to the chemical and process industries. Theoretical courses consist of the solution and discussion of design problems dealing with portions of manufacturing processes. These courses are supplemented by laboratory courses designed to emphasize the fundamental principles of chemical engineering by experiments on pilot plant size equipment." Lawrence Hein, assistant professor, taught all chemical engineering courses that year.

The following year, Professor Henry T. Ward and Instructor Wickliffe Skinner reworked the curriculum. The 1947–1948 curriculum looked like this:

Sophomore Year

Fall	Spring
Chem. 214, Qualitative Analysis	Ch.E. 262, Chemical Engr. Fundamentals
Draw. 212, Mechanical Drawing	Chem. 254, Quantitative Analysis
Engl. 322, Composition for Engineers	Speech 202, Public Speaking
Math. 214a, Differential Calculus	Math. 214b, Integral Calculus
Phys. 134a, Engineering Physics	Phys. 134b, Engineering Physics
Military Art 201a	Military Art 201b

Junior Year

Fall	Spring
Ch.E. 314, Industrial Stoichiometry	Ch.E. 323, Unit Operations
Chem. 315a, Organic Chemistry	Ch.E. 331, Unit Operations Lab
Chem. 403a, Physical Chemistry	Chem. 315b, Organic Chemistry
Chem. 401a, Physical Chemistry Lab	Chem. 403b, Physical Chemistry
M.E. 305, Mechanics	Chem. 401b, Physical Chemistry Lab
	C.E. 315, Strength of Materials

Senior Year

Fall	Spring
Ch.E. 403, Unit Operations, contd.	Ch.E. 412, Unit Operations, contd.
Ch.E. 421, Seminar	Ch.E. 423, Plant Design
Ch.E. 433a, Thermodynamics	Ch.E. 424, Physical Metallurgy
Ch.E. 432, Unit Operations Lab	Ch.E. 433b, Thermodynamics
Ch.E. 422, Plant Design	Ch.E. 441, Unit Operations Lab
Econ. 203a, Principles of Economics	Ch.E. 452, Organic Technology
E.E. 363, 311, Electrical Engineering	Econ. 203b, Principles of Economics

With the exception of Physical Metallurgy, which was taught by John E. Shoemaker of the Ordark Research Department, Ward and Skinner taught all of the chemical engineering courses. They shared responsibilities for the Unit Operations courses, which covered fluid flow, heat transfer, evaporation, diffusional processes, humidification, distillation, absorption, extraction, crystallization, filtration, mechanical separation, crushing, and grinding.

The Barn. Located just left of center and to the north of the Fine Arts Building. (Mullins Library Special Collections, photo collection #941)

Student chemical engineers took their classes in a variety of locations. The Chemistry Building still housed most of their chemistry and chemical engineering classes. But by the end of this transition era, most of their chemical engineering classes had been moved to friendlier confines. In 1947, the Federal Works Administration constructed seven temporary wooden buildings on the University of Arkansas campus. One of these buildings, known as "the Barn," housed chemical engineering laboratories. It was located north of Razorback Hall and east of Garland Street, near where the Fine Arts Center is located today.

Of the students taking their classes in "the Barn," one stood out: Dana Joyce Jesswein. Jesswein, the first woman to earn an engineering degree from the U of A, graduated in 1948 with her B.S.Ch.E. Jesswein enjoyed her time at the university and later recalled, "life was great for me in engineering school—I was taking courses I liked (with the exception of chemistry lab). I got along well with fellow students, made many good friends[,] and studied with them. Although there was probably some resentment, I was never aware of any at all. Outside engineering school, occasionally someone would say, 'They'll never let a female graduate from engineering,' but I just laughed." She also remembered that one of her professors did have a problem, but it was more his than hers:

> There is one humorous incident I remember, but I can't recall the instructor's name. I had taken two of the shop courses required—machine shop and woodworking—which I enjoyed very much. One of the instructors also taught forge and foundry, which I was enrolled in for the upcoming semester. I received a notice to report to Dean Stocker's office, and immediately began to worry about what I had done to get in trouble. Dean Stocker told me that this instructor was accustomed to using rather profane language in class, which he did not feel was appropriate in front of a female, but trying to "abstain" put a tremendous amount of stress on him. I assured Dean Stocker that he could feel free to use as much profanity as he wished, and it wouldn't bother me in the least. Dean Stocker said it might not bother me, but it bothered the instructor, so would I mind taking any course I wished as a substitute? So I took flying lessons instead.

Jesswein paved the way for future women engineers of all fields. In 1950, the first women to earn bachelor's degrees in civil engineering and electrical engineering graduated, and seven years later a woman received a bachelor's in mechanical engineering. Jesswein and these other early graduates opened the door for more and more women to enroll in engineering courses. In the forties and fifties, administrators, parents, teachers, and other students might have looked askance at women entering a so-called man's field, but today, women engineers draw little notice.

The transition era also saw a number of developments designed to foster research among chemical engineering faculty members and, in some instances, students. On July 1, 1943, the University of Arkansas Board of Trustees established a Bureau of Research "to determine the opportunities for further industrial development in Arkansas." The bureau consisted of research and administrative staff at the university in Fayetteville and at a research division in Little Rock. Among the staff at Fayetteville were Professor of Chemical Engineering William L. Belvin and Emeritus Professor of Chemistry Harrison Hale. The bureau worked with various departments on campus and published research bulletins, which were available free to citizens of Arkansas. It was originally housed in two temporary wooden buildings just south of Dickson Street and east of Buchanan; a $60,000 pilot plant and laboratory was added in March 1946.

The bureau opened up research possibilities in many different realms, but it especially focused on Arkansas's "natural resources, possibilities and methods of processing, and marketing outlets." Some of its ventures dealt directly with chemical engineering, including studies of the "processing, fabrication, or manufacture of basic raw materials produced in Arkansas, . . . the quantity and quality of water available for industrial use, . . . [and] the discovery of new and improved industrial processes partly through the operation of pilot plant laboratories." The Bureau of Research made its findings available to the Arkansas Resources and Development Commission.

Another new research outlet for U of A faculty was created when the U.S. Army Office of the Chief of Ordinance created the Ordark Research Department at the University of Arkansas in 1945. Ordark reflected the federal government's growing interest in basic scientific research during and after World War II. According to the *University Catalog,* "the subject of the research program is concerned with fundamental scientific research, the findings of which should have a direct bearing on improvements of materials and supplies used by the Ordinance Department; it is also hoped that industrial applications may be derived from this research program."

Professor of Chemistry Edgar Wertheim served as the chairman of the Ordark Advisory Committee, and John Shoemaker, who taught Physical Metallurgy as part of the chemical engineering curriculum, was assistant director of the department. Although he did not teach in the chemical engineering department, Wladimir W. Grigorieff, the director of Ordark, was also a trained chemical engineer.

The Ordark Research Department built a "new and modern" building (now home to the geology department) directly north across Dickson Street from the Bureau of Research buildings. Ordark also offered graduate assistantships and fellowships, worth up to fifteen hundred dollars for a twelve-month term, to students in scientific and engineering fields, including chemical engineering. Its work was paid for by the army and was classified as secret.

W. W. Grigorieff. (Simpson, 178)

In early 1948, the Institute of Science and Technology was created at the University of Arkansas to coordinate a broad research program in conjunction with the university's academic program. W. W. Grigorieff, director of Ordark, became the new institute's director. According to the institute's mission statement, its goal was to "develop an expanded program of training in science and technology on the senior college and graduate levels . . .[,] to coordinate and promote fundamental research at the University in all sciences . . .[,] to conduct applied research on a contractual basis for individuals, industries, or government agencies . . .[, and] to work in close cooperation with public and private research agencies in this region and with organizations engaged in the industrial and agricultural development of the state." Ordark and the Bureau of Research became components of the Institute of Science and

Technology, and, as a result of the creation of the institute, the U of A became a member of the Oak Ridge Institute of Nuclear Studies.

While the new research programs beckoned to faculty members, chemical engineering students availed themselves of new social and extracurricular activities. During the fall semester of 1938, a number of engineering students met together and came up with the idea of a cooperative housing venture. After numerous meetings, the group christened itself the Engineers' Cooperative Housing Organization (ECHO). According to its constitution, "the purpose of the organization shall be to secure for young men desiring to obtain a college education the advantages of cooperative efforts in providing room and board; to foster a spirit of fellowship; and to promote social activities within the group and on campus."

The thirty-six engineering students who lived in the ECHO house paid a reasonable rate—$16.50 per month—for room and board. The house itself was located on the corner of Arkansas Avenue and Maple Street. The endeavor proved to be so popular that the organization added

ECHO exterior. (*Razorback,* 1940, 121)

ECHO interior. (*Razorback,* 1944, 99)

an ECHO Annex, located across Maple Street from the original house, and increased membership to fifty-five after one year. The ECHO Annex lasted only one year, though; by the fall of 1941, ECHO numbers had returned to thirty-six. Membership in the organization was open to nonengineering students (they could make up to 25 percent of the total) who showed financial need and dedication to their schoolwork. "It matters not how brilliant a man may be, or how high he may stand in his class," noted ECHO president George Doerries in 1940, "if he does not possess a reasonable amount of congeniality and personality, and does not show some financial need it is not likely that he will be admitted into the organization."

Besides the obvious benefits of cheaper cost of living and shared concerns over the engineering curriculum, the ECHO house proved to be a useful social venture. Although only serious and dedicated students made it onto the ECHO membership roster, its organizers realized that success after college resulted from more than just "book learning." "Advancement by reason of education will depend upon the number and character of contacts that are made with people," Doerries noted. "Even the most brilliant ideas are useless if the one originating them cannot get associates to accept them. This is one of the things the ECHO House is attempting to accomplish." Social events usually took the form of parties and picnics, although plans did not always work out, as the *Arkansas Engineer* reported: "The ECHO lads are thinking of going into business as rainmakers. Technique—simple. They plan a picnic and the deluge descends upon the appointed day."

Theta Tau Engineering Fraternity, which had been established at the U of A in 1928, also played a role in the daily lives of many chemical engineering students during the 1938–1948 period, as one former student remembered: "This [Theta Tau] was very important on campus in those days, [it] even had a social influence on campus. We had our own house and paid $25 room and board. We had our house manager and ate steaks the first day of the month and peanut butter and jelly sandwiches for 29 days; with money left over for a few 'beer busts.'"

Engineers' Day, 1940. *l. to r., front row,* Guard Louie W. Walter; St. Patricia Dorothy Aday; St. Patrick, William M. Hathaway, BSChE 1941; Guard Chester Doty; *back row,* Pages Gordon Wittenberg and Bob Wetzel. (*Razorback,* 1940)

Engineering students also continued to plan and participate in Engineers' Day in March of every school year, and they added new features regularly. For instance, in 1939, the engineers for the first time erected a ten-foot shamrock between Old Main's towers. Between 1942 and 1945, Engineers' Day was celebrated in late January instead of the customary St. Patrick's Day due to the term system instituted by the university.

Continuities, however, overshadowed the changes made to Engineers' Day. The rivalry between engineering and agriculture students continued to spark confrontations between the two, much to the consternation of university administrators and townspeople. In 1942, the engineers succeeded in painting some of the Agris' prize possessions, their award-winning cattle, "with shamrocks from head to foot." The rivalry only intensified after the war. The *Arkansas Engineer* noted the following fallout in one editorial:

Results of Engineer's Day:

1. A few cases for the infirmary, both Agri and Engineer, consisting of minor cuts and bruises.
2. A few impromptu haircuts.
3. Green shamrocks covering the campus and store windows up town.
4. A green calcimine shamrock on the side of an Organized House for Agri's.
5. A broken window in the O.H.A.
6. Considerable green calcimine spilled on the lawn and street facing the O.H.A.
7. Payment of claims, by the Engineering Council, for broken street light and defaced traffic light.

Results of Agri Day:

1. No cases for the infirmary.
2. An impromptu haircut, administered to a Traveler reporter who is a non-engineering student.
3. A few white feet covering the campus and store windows.
4. A white calcimine foot on the front of an Organized House for Engineers
5. Very little white calcimine spilled on the lawn and street facing the O.H.E.
6. White calcimine sprayed on the O.H.E. from top to bottom, including the front and half-wayaround one side.

Although the results of Engineers' Day seem more dramatic at first glance, the final item on the list resulted in a damages claim being filed by the owner of the ECHO house.

Dean of Engineering George Stocker created a special committee made up of representatives of both the Colleges of Engineering and Agriculture following the 1947 festivities. Stocker's goal was to continue both colleges' celebrations without the wanton destruction of previous years. The committee met and formulated a resolution, the gist of which the *Arkansas Engineer* reported: "The College of Engineering and the College of Agriculture, both, agree not to interfere in any way whatsoever with the observance carried on by the other college. Painting of emblems will be done with water-mix calcimine and will be confined to streets and sidewalks. No houses or windows will be painted.

Furthermore, if the agreement is broken and damage results, the Engineering Council or the Agri Day Association, as the case may be, will pay all damage resulting."

Although this resolution addressed Stocker's concerns, an editorial from the same issue of the *Arkansas Engineer* foreshadowed another dilemma. "Engineer's Day must [be] participated in by more students if it is to survive," it noted. "The banquet last year was attended by approximately one-fourth of the students enrolled in engineering; the breakfast was attended by one-tenth; the fireworks display was attended by one-twelfth; convocation brought forth even smaller attendance." Just as ominous, calls began to be heard for combining Engineers' Day with other campus festivities, including Commerce Day, Lawyers' Day—even Agri Day. The resulting hodgepodge celebration would then be subsumed under the recently established Gaebale function. As Engineers' Day neared its fortieth anniversary, it seemed to be nearing its end also.

Chemical engineering students continued to take part in the activities of Alpha Chi Sigma. Like most other organizations, the professional chemistry fraternity watched its membership dwindle during World War II, but the group always maintained at least a modest level of activity. For the period 1939–1948, Alpha Chi Sigma held regular meetings during the school year for members to discuss pledges, initiation ceremonies, and finances and to listen to guest speakers. Entertainment at the regular meetings featured renditions of Alpha Chi Sigma songs, games, quiz contests, puzzle solving, and even a "chemical track meet." Twice a year, the group held banquets. The Laboratory Banquet, held on campus at the Chemistry Building, introduced new initiates and potential pledges to the organization; the semiformal Founders' Banquet—usually held off campus at locales such as the Mountain Inn, the Washington Hotel, the Fayetteville Country Club, or the Campus Grill—allowed members to show off their organization to dates. Smokers, usually held at a faculty member's home near the beginning of each semester, were designed to acquaint new university students with the fraternity.

Alpha Chi Sigma did not concern itself only with social events, however. The group sponsored high school chemistry contests, paid the expenses of guest speakers while they visited the U of A, and offered tutoring services for students struggling with their chemistry class work. Each year the organization gave a membership into the American Chemical Society to a top graduating chemist or chemical engineer.

War dominated the concerns of all students and faculty for most of the 1938–1948 period, and campus life revolved around the war effort. "Total war is not confined to the battlefront, the factory, and the home," wrote U of A president Arthur M. Harding in December 1942. "It must be waged in every college and in every university in America. Every professor and every student has his part to play."

The university's administration changed the academic calendar once again to decrease the time it took to receive a diploma. Besides the changes made from the semester to the term system, summer school offerings were expanded as well. Due to the increase in course offerings, and the fact that more than a hundred U of A faculty members served in active duty, the university had to make a multitude of temporary hirings, as is evident from the tenures of teachers in the chemistry and chemical engineering departments.

The war brought immediate changes to the university's student population, also. "The college year of 1942–43 saw 148 students in civilian pilot training, 156 in the pre-radar unit, and 1,200 in the army aircorps unit, bringing the total resident enrollment to 3,786, compared with the civilian student enrollment of 2,466 the preceding year," noted university historian Harrison Hale. "The majority of male students were in army uniform."

The influx of military trainees led to changes in the university's physical environment. Camp Neil Martin, named after the first U of A alumnus to die in the war, was built to the north and east of the football stadium on Garland Avenue and Cleveland Street. It housed some six hundred young men in training for the Army Air Corps, and included a cafeteria capable of holding seven hundred and fifty. Razorback Hall (now Gibson Hall), a men's dormitory, and Mary Anne Davis Hall, a cooperative home for women, became barracks for members of the Army Specialized Training Program. Later, more barracks were created in various camps along the edge of campus: Terry Village, Lloyd Halls, and Camp Leroy Pond all housed hundreds of newcomers. After the war, the camps were remodeled into housing for veterans and married students.

The U of A also participated in the war effort by training individuals for work on the home front. Even before the United States became an active belligerent, the university launched the Engineering, Science, Management Defense Training program jointly with the federal government. After December 1941, the program became known as the Engineering, Science, Management War Training program (ESMWT).

The program trained men and women around Arkansas "in technical skills for work in our country's defense industries." Most of the instruction occurred off campus, with university faculty training workers on topics such as the "Chemistry of Powder and Explosives" to prepare them for work in munitions plants. Approximately twenty-five hundred Arkansans participated in the ESMWT program.

For university students who did not serve in the military, school life continued. Excepting a few upper-level undergraduate and graduate-level courses, the university continued to offer a full range of classes. "A full course of instruction—in the arts, the sciences, and the professions—was maintained for our civilian students," wrote U of A president Harding. "Wherever necessary, our civilian courses were adjusted to meet the needs of the time, and in all cases the latest scientific, political, economic[,] and social developments were included in the instruction."

Athletics, although subdued by the exigencies of war, continued to be a center of campus attention. In 1943, the Board of Trustees created a permanent athletic committee, composed of one trustee, one alumnus, and one outside member. Two more alumni were added to the board in suc-

Campus during World War II. (*Razorback,* 1944, 46–47)

ceeding years, and, in 1945, the U of A acted upon the committee's recommendation and hired John Barnhill as the university's first athletic director. Overall, however, the university's athletics, like so much else on campus, were in a state of flux during the war.

Coach Fred C. Thomsen left the football team to join the army in 1942; three coaches led the team onto the gridiron over the next three years. The football Razorbacks compiled a 13–26–1 record during the war. When Barnhill arrived as athletic director in 1946, he took over head football coaching duties and led the team to a series of successful seasons, culminating in back-to-back bowl games in 1947 and 1948. Coach Glen Rose's basketball teams continued to be successful; in 1941 the team posted an undefeated conference mark. In 1942, Rose left for the army and was replaced by E. W. "Gene" Lambert, who continued Rose's successful ways. The 1944 team tied for the Southwestern Conference championship.

Many residents of Fayetteville volunteered for service during World War II, and many others moved to California in search of higher-paying jobs. Those that remained worked extra hours at the local lumber mills and canning plants, bought war bonds, and ate home-grown food—all part of the war effort. When the war ended, the city experienced a period of economic expansion; by the late 1940s, more than two hundred retail and wholesale outlets had located themselves within Fayetteville's confines. New eateries included Ralph Ferguson's Blue Mill Café on Block Street. Maple trees lined College Avenue, and more and more gas stations and restaurants popped up along the thoroughfare as the city gradually crept northward.

Fayetteville continued to grow throughout the 1950s, and the chemical engineering program did, too. Under the steady hand of Colonel Maurice Barker, who arrived at the Department of Chemical Engineering in the fall of 1948, the period of development and foundation-laying ended, and a new period of maturation began.

Biographical Sketches, 1938–1948

Stuart "Stu" McClain arrived on the University of Arkansas campus in 1938 as an instructor of chemistry. McClain was the first professor at the university with a degree in chemical engineering, having earned his B.S.Ch.E. (1929), M.S.Ch.E. (1930), and Ph.D. (1933) from his home state school, the University of Michigan. While in college, he worked as an assistant in Gas and Fuel Analysis and received a Michigan Gas Research Fellowship and memberships in the Junior Research Club and Sigma Xi. McClain's arrival prompted the move toward separating chemical engineering from the Department of Chemistry and establishing it as a separate unit.

Before arriving at Arkansas, McClain worked as a chemist with Fire Test and Textile Development, as chemist for the City of Detroit, and as assistant professor of chemistry at Michigan. He held professional memberships in Alpha Chi Sigma, the American Chemical Society, the American Institute of Chemical Engineers, the Society for Promotion of Engineering Education, and the Engineering Society of Detroit. McClain also was a member of the Reserve Army Corps.

Stuart McClain. (*Arkansas Engineer,* November 1938, 18)

The *Arkansas Engineer* described McClain as a "capable, pleasantly-serious and conscientious fellow interested in his work and his fellow workers. He's not a spectator, and he would rather play his own tennis game than watch a football game. He likes bridge, plays handball and tennis, and does a lot of swimming. . . . A serious, hardworking, well developed fellow—he doesn't like swing music and most radio programs—but prefers symphonies and history."

McLain received a promotion

to assistant professor of chemistry in 1940, but in May of 1941 the *Arkansas Engineer* noted that "the Chemicals are losing their Dean of Assumptive Analysis, Dr. McLain, who will probably serve a year in the army," adding that "it will take two men and a span of mules to replace Dr. McLain." McLain ended up serving in the military until the end of World War II. Following the war, he obtained a position at Yale University, where he eventually became chairman of the chemical engineering department.

William L. Belvin arrived at the University of Arkansas in 1944 as the university's first full professor of chemical engineering. He was a staff member of the Bureau of Research, where his research centered on cottonseed oil. Belvin received both his B.S.Ch.E. and M.S.Ch.E. from North Carolina State College.

Professor Belvin is significant for being the first full professor of chemical engineering at Arkansas, but his duties for the Bureau of Research kept him from teaching during the four years he was on staff. He resigned from the University of Arkansas on November 15, 1948.

After leaving the U of A, Belvin became senior research chemist with Allied Chemical and Dye Corporation. He then served as director of the Herty Foundation of Savannah, Georgia. He retired from Herty in August 1980.

Henry T. Ward and *Wickliffe Skinner* both came to the University of Arkansas in 1947. Ward became professor of chemical engineering and the first head of the Department of Chemical Engineering; he had received his B.S.Ch.E. from the University of Michigan, a M.S. from the University of Wyoming, and a Ph.D. from the University of Michigan. "Wick" Skinner had earned a B.S.Ch.E. from the University of Oklahoma and was named instructor in chemical engineering.

Ward and Skinner taught all chemical engineering courses for the 1947–1948 school year, including graduate-level classes. Both left the University of Arkansas after one year. Ward became head of the Department of Chemical Engineering at Kansas State University, and Skinner left for the Midwest Research Institute.

Ward, whom a former colleague remembered as gregarious and friendly yet plainspoken and straightforward, guided Kansas State's chemical engineering program to maturity during the 1950s in much the same way that Colonel Maurice Barker guided Arkansas's. In 1960, he died in a bizarre accident, as Professor Charles Oxford recalled: "Henry Ward was deathly afraid of airplanes; in fact, he wouldn't fly. He took a train from where he was in Kansas going to Chicago, and at a crossing a truck ran into the Pullman car in which [he] was riding and he was killed." Ward was the only person involved in the accident to die.

Peter N. Bragg Jr. graduated from the University of Arkansas in 1942 with a B.S.Ch.E. The following year, he joined the university faculty as an instructor in chemistry, a position he held for one year. In June 1944, Bragg began work with the Naval Research Laboratory. His goal there: the production of weapons-grade uranium as part of the Manhattan Project, the U.S. military's massive scientific undertaking to create an atomic bomb.

Bragg worked at the Philadelphia Navy Yard on a secret pilot plant. There, on September 2, 1944, a cylinder of lethal gas exploded, killing Bragg and fellow civilian worker Douglas Meigs. Because of the deep secrecy of the Manhattan Project, the circumstances of Bragg's death did not become known until well after the fact. Even after the government made known details of the project, his family did not learn exactly how he died.

Peter Newport Bragg. (*Razorback*, 1943, 40)

In the early 1990s, Arnold Kramish, who witnessed the accident, and Braxton Bragg, Peter's brother, began an effort to gain official recognition of Peter Bragg's sacrifice. On June 21, 1993, Bragg posthumously received the Navy Meritorious Civilian Service Award, the highest award given to a civilian employee of the Navy Department.

Bragg received the first M.S.Ch.E. granted by the University of Arkansas. It was awarded posthumously in 1945.

Walter S. Dyer became instructor in chemistry in 1929. He served in that capacity until 1936, when he took a two-year leave of absence to complete his doctoral studies. In 1938 Dyer returned to Arkansas as assistant professor of chemistry. He became associate professor in 1945.

Walter S. Dyer. (*Arkansas Engineer*, November 1938, 18)

Born in Russellville, Dyer moved as a young boy to Fayetteville, where he attended the University of Arkansas, receiving his B.S. in chemistry with honors in 1924. He earned his M.S. (1925) and Ph.D. (1938) from the University of Minnesota.

Dyer also served as a teaching assistant with the University of Minnesota and as professor and head of sciences at Union College (Kentucky). He held memberships in Sigma Xi, Phi Lambda Upsilon, Alpha Chi Sigma, and the American Chemical Society. The *Arkansas Engineer* wrote of him that "he seems to have decided on a career in Chemistry because it seemed to fulfill his desire for usefulness, exactness, and logic." He died in 1945.

Chapter Four

The Barker Years, 1948–1961

The late 1940s and the 1950s are remembered as a period of overall prosperity for the United States. This prosperity filtered into colleges and universities in the form of increased enrollments, thanks in large part to the GI Bill. But prosperity is not always positive. As the number of college students rapidly expanded following World War II, faculties and administrations faced overcrowding in classrooms and dormitories, problems only compounded by the shortage of qualified teachers in all fields.

The U of A was not immune to the problems created by rapid enrollment increases. By 1948, 5,626 resident students called the university their academic home. This number more than doubled the pre–World War II peak enrollment; enrollment in summer sessions doubled as well. The increased enrollments and shortage of teachers would be reflected in the Department of Chemical Engineering. In 1955, 119 students were enrolled in the chemical engineering program (38 freshmen, 40 sophomores, 21 juniors, and 20 seniors). Thirty-two seniors graduated with B.S.Ch.E.s in 1956—compared to peaks of nine during the Hale years and fifteen during the transition years. Despite the growth in student numbers, the majority of faculty members between 1948 and 1960 stayed only a handful of years. But a few stayed longer, and they provided an enduring foundation for the future.

At the same time as they faced increases in enrollment and shortages of faculty, American colleges uncovered a new source of financial support: research projects paid for with public and private funds. The cold war between the United States and the Soviet Union that developed after the end of World War II created an arms and technology race that mushroomed in the 1950s and 1960s and brought renewed emphasis on science and engineering research and education. Also, the prosperity of the postwar period relied on industrial growth, and industrial funding of research projects at universities became increasingly common. These new sources of funding led to increasing emphasis on coordinating research activities through entities such as the Engineering Experiment Station and the Institute of Science and Technology.

This emphasis on research projects resulted in an interest in expanding and improving the U of A Graduate School, an endeavor which reflected the state's need for highly educated, specialized professionals in a number of areas—including business research and the oil, mining, and chemical industries. To achieve the goal of producing such professionals, and of ending the state's reliance on trained workers from the outside, the university decided to, first, upgrade its existing master's programs, and, second, create doctoral programs as soon as possible. The university's first Ph.D. (in humanities, social sciences, physical sciences, or medical sciences) and Ed.D. programs were offered in 1950. The Department of Chemical Engineering reflected the emphasis on graduate studies when it produced its first earned M.S.Ch.E.s in 1951.

Another major change came in 1948, when Silas Hunt of Texarkana became the first African American student to enroll at the University of Arkansas since 1872. Hunt entered the School of Law during the spring semester and resigned due to illness that summer—he died of tuberculosis the following April. While at the university, Hunt studied in an almost completely segregated environment, but his tenure at the school provided an example for other blacks who wished to attend the U of A. The Law School slowly allowed larger numbers of African Americans to enroll, and, as the 1950s wore on, barriers at other colleges and departments fell as well. By the end of the decade, hundreds of African Americans were enrolled at the university.

Although the climate on the U of A campus appeared less volatile than on other southern universities during the era of integration, some conflict inevitably occurred. Besides tensions between white and black students, troubles erupted when, in 1958, the Arkansas Legislature passed Act 10, which required all faculty members to submit annual lists of every organization to which they had contributed money or which they had been members of for the previous five years. Act 10 had been sponsored by segregationist legislators in the wake of the Little Rock Central High School debacle as a means of rooting out any potential integrationist troublemakers—members of the American Civil Liberties Union, the Urban League, or the American Association of University Professors, for instance—and reflected the tensions inherent in integration. The Arkansas Supreme Court upheld the act, but the United States Supreme Court declared it unconstitutional in 1960. The effects of Act 10 lingered into the 1960s, however, and the American Association of University Professors placed the University of Arkansas on its "List of Censured Administrations" for four years. The act was but one more reason for the fluidity of faculty at the university.

The faculty of the U of A Department of Chemical Engineering exhibited a notable fluidity between 1948 and 1960. Behind this fluidity lay a fundamental problem faced by most engineering colleges in the United States: this time period witnessed a shortage of trained engineers in all fields at the same time that industrial expansion created high-paying jobs in the private sector. Adding to this difficult circumstance, the University of Arkansas as a whole did not have a competitive pay scale for faculty as compared to other universities, and thus lost teachers to other schools at an alarming rate. These situations created the fluidity, the coming-and-going, that is so evident in the following list. Two professors, Colonel Maurice Barker and Dr. Charles Oxford, proved to be mainstays of the department, however, and another long-term addition, Dr. James Couper, came on board toward the end of the Barker era.

1. Maurice E. Barker. Professor of Chemical Engineer and Head of the Department (1948–1960); Emeritus Professor of Chemical Engineering (1960–1979). B.S., Valparaiso University; A.M., University of South Carolina; Sc.D., Massachusetts Institute of Technology.

2. Harry Scholars Autrey. Instructor in Chemical Engineering (1948–1954). B.S.Ch.E., University of Arkansas.

3. Vincent Vinett Drewry. Instructor in Chemical Engineering (1946, 1949–1952). B.S.Ch.E., University of Arkansas.

4. Carl Mock Gamel Jr. Assistant Professor of Chemical Engineering (1948–1950). B.S.Ch.E., University of Arkansas; M.S., Massachusetts Institute of Technology.

5. Edmund Deberry Lilly. Instructor in Chemical Engineering (1948–1949). B.S.Ch.E., University of Arkansas.

6. Richard Matthaei. Instructor in Chemical and Civil Engineering (1948–1949); Instructor in Industrial Engineering (1949–1950). B.S., New Mexico School of Mines.

7. Charles William Oxford. Assistant Professor of Chemical Engineering

(1948–1957); Professor of Chemical Engineering (1957–1988); Associate Dean of Engineering and Associate Director of the Engineering Experiment Station (1960–1968); Administrative Vice President of the University (1968–1980); Vice President for Academic Affairs (1980–1988); Emeritus Professor of Chemical Engineering (1988–). B.S.Ch.E., University of Arkansas; Ph.D., University of Oklahoma; P.E.

8. John Milton Lenoir. Assistant Professor of Chemical Engineering (1949–1952). B.S., University of Illinois; M.S., University of Iowa; Ph.D., University of Illinois.

9. John Paul Sanders. Instructor in Chemical Engineering (1950–1954); Associate Professor of Chemical Engineering (1957–1965). B.S.Ch.E., M.S.Ch.E., University of Arkansas; Ph.D., Georgia Institute of Technology; P.E.

10. William Joseph Smothers. Assistant Professor of Ceramics (1950–1951); Associate Professor (1951–1954). B.S., M.S., Missouri School of Mines; Ph.D., University of Missouri.

11. Larry Gordon Sloan. Instructor in Chemical Engineering (1954–1955). B.S., University of Arkansas.

12. James Edward Stice. Assistant Professor of Chemical Engineering (1954–1956). Associate Professor of Chemical Engineering (1962–1967); Professor of Chemical Engineering (1967–1968). B.S.Ch.E, University of Arkansas; M.S.Ch.E., Ph.D., Illinois Institute of Technology.

13. Maurice Kendall Testerman. Associate Professor of Chemical Engineering (1951, 1954–1959). B.S., M.S., Ph.D., Virginia Polytechnic Institute.

14. Jewel G. Rainwater. Instructor in Chemical Engineering (1957–1960); Assistant Professor of Chemical Engineering (1960–1965). B.S.Ch.E., M.S.Ch.E., University of Arkansas.

15. Bill Harrell. Instructor in Chemical Engineering (1956–1957). B.S.Ch.E., University of Arkansas.

16. William A. Myers. Instructor in Chemical Engineering (1956–1957). B.S.Ch.E., University of Arkansas. For more on Myers, see chapter 7.

17. Tom Leland. Assistant Professor of Chemical Engineering (1953–1954). B.S.Ch.E., Texas A & M; M.S.Ch.E., University of Michigan; Ph.D., University of Texas.

18. James Riley Couper. Associate Professor of Chemical Engineering (1959–1965); Professor of Chemical Engineering (1965–1989); Head of the Department of Chemical Engineering (1969–1979); Professor Emeritus of Chemical Engineering (1989). B.S.Ch.E., M.S.Ch.E., D.Sc., Washington University; P.E.

As the faculty development reflected a national trend—fluidity—so too did the chemical engineering curriculum at the U of A. Engineering colleges and departments around the nation developed curricula based on a uniform freshman year followed by three years of more complex work in a student's specialized field of study. In accordance with recommendations from the Engineers' Council for Professional Development, engineering curricula also began to emphasize outside, nonengineering fields—such as economics, management, humanities, and social studies—in order to develop more well-rounded students and professionals. Concurrent with these changes, graduate study came to be emphasized as a means of creating more in-depth and specialized understanding of engineering problems. The Department of Chemical Engineering at the U of A, under Colonel Barker's leadership, implemented curricular changes in accordance with these developments.

The first year of Colonel Barker's tenure brought many changes to the undergraduate chemical engineering curriculum. These changes reflected both Barker's emphasis on practical experience and chemical engineering's refinement as a profession. The curriculum for students entering chemical engineering in 1948–1949 was as follows:

Sophomore Year

Fall	Spring
Phys. 134a, Engineering Physics	Phys. 134b, Engineering Physics
Phys. 141a, Laboratory Physics	Phys. 141b, Laboratory Physics
Math. 214a, Differential Calculus	Math, 214b, Integral Calculus
I.E. 212, Engineering Drawing	E.M. 203, Statics
Chem. 254, Quantitative Analysis	Ch.E. 213, Industrial Stoichiometry

Ch.E. 262, Fundamentals & Materials	Ch.E. 201, Fuel & Gas Analysis Lab
M.A., 201a, 2nd Year Basic	M.A. 201b, 2nd Year Basic

Junior Year

Fall	**Spring**
Ch.E. 323a, Unit Operations	Ch.E. 323b, Unit Operations
Ch.E. 332a, Unit Operations Lab	Ch.E. 332b, Unit Operations Lab
Chem. 315a, Organic Chemistry	Chem. 315b, Organic Chemistry
Chem. 403a, Physical Chemistry	Chem. 403b, Physical Chemistry
Chem. 401a, Physical Chemistry Lab	Chem. 401b, Physical Chemistry Lab
E.M. 303, Dynamics	E.M., 314, Mechanics of Materials
Spe. 202, Public Speaking	

Senior Year

Fall	**Spring**
Ch.E. 422, Plant Design	Ch.E. 423, Plant Design
Ch.E. 433, Thermodynamics	Ch.E. 413, Metals Corrosion
Ch.E. 403, Organic Technology	Ch.E. 463, Projects
Ch.E. 441, Chemical Engineering Seminar	E.E. 302, D & A/C Apparatus
E.E. 323, D & A/C Circuits	E.E. 311, Electrical Equipment Laboratory
Econ. 203a, Principles of Economics	Econ. 203b, Principles of Economics
Humanities Elective	Humanities Elective

The three-semester course Unit Operations that had been offered the previous year became a two-semester offering, and the courses Dynamics and Mechanics of Materials were added. The department dropped Physical Metallurgy as a required course and replaced it with Metals Corrosion. Other new additions included the two Electrical Engineering classes and one lab during the senior year, and Chemical Engineering 463, Projects, which focused on "problems of economic balance in planning, installing, and operating chemical plants, and process equipment; the preparation of finished reports upon engineering subjects; and market analysis." In order to graduate, chemical engineers had to complete 145 semester hours.

By the 1950–1951 school year the faculty had devised a new, more varied curriculum for students who wished to specialize in one of three subfields of chemical engineering. Option one, known as the Regular Course, was the coursework as devised in 1948. Option two, the Petroleum Option, reflected the petroleum industry's growing need for professionally trained chemical engineers. The Petroleum Option followed the regular course, for the most part, replacing Ch.E. 412 with Ch.E. 522, Petroleum Technology, and two hours of humanities with Ch.E. 492, Fluid Flow. Students in the Petroleum Option also handled assignments dealing with petroleum refining in Ch.E. 422, Plant Design. Option three, the Wood Products Option, also reflected demand for trained chemical engineers in a specialized industry. For the Wood Products Option, Ch.E. 482, Wood Technology, replaced Ch.E. 412, two hours of work in Ch.E. 506, Research, replaced the same amount of humanities, and a wood technology design problem had to be solved in Ch.E. 422.

By 1956–1957, the chemical engineering curriculum had undergone even more changes. Notably, the Petroleum Option and Wood Products Option had been dropped in favor of a single curriculum. Steps also had been taken to incorporate more humanities offerings into the student work load; new requirements included English 331, American Literature, History 201, The U.S. as a World Power, and Western Civilization 101, Institutions and Ideas, which could be exchanged for a two-semester Fine Arts course or Psychology and Sociology. Physics 361, Elements of Nuclear Physics, was added to the junior year.

Until the end of the Barker years, the chemical engineering faculty continued to tinker with the curriculum. The English, history, and Western Civilization requirements were replaced with more general two-semester humanities electives. In 1958, Ch.E. 473, Elements of Nuclear Engineering, was authorized as a substitute for Physics 361, Elements of Atomic Physics. In 1959, Physics 361 was required once again. In 1959, a third semester of Unit Operations again became part of the curriculum, and Mathematics 340, Differential Equations, and Ch.E. 403, Chemical Processes were added. That same year, Ch.E. 412, Materials of Construction, became an elective; students could take it or Technical Administration, Reactor Design and Operation, or Petroleum Processing. By the end of the Barker years, chemical engineering students at the U of A could expect to take the following courses:

Freshman Year

Fall	Spring
English 101, Composition	Engl. 102, Composition
Chem. 104, General Chemistry	Chem. 104, General Chemistry
I. E. 112, Engineering Graphics	Math. 130, Analytic Geometry
G. E. 101, Freshman Orientation	Math. 251, Calculus I
Math. 128, College Algebra and Trigonometry	I.E. 102, Engineering Problems
Humanities Elective	I.E. 113, Engineering Graphics II
Military or Air Science	Military or Air Science

Sophomore Year

Fall	Spring
Phys. 207, Engineering Physics	Phys. 207, Engineering Physics
Phys. 208, Laboratory Physics	Phys. 208, Laboratory Physics
Math. 252, Calculus II	Math. 253, Calculus III
Econ. 203, Principles	E.M. 200, Statics
Chem. 222, Quantitative Analysis	Ch.E. 213, Ind. Stoich. & Elem. Thermo.
Ch.E. 201, Fundamentals	Econ. 203, Principles
Ch.E. 201, Fuel and Gas Analysis Lab	Military or Air Science
Military or Air Science	

Junior Year

Fall	Spring
Ch.E. 323, Unit Operations I	Ch.E. 324, Unit Operation II
Chem. 364, Organic Chemistry	Ch.E. 332, Unit Operations Lab I
Chem. 342, Physical Chemistry	Chem. 364, Organic Chemistry
Chem. 343, Physical Chemistry Lab	Chem. 342, Physical Chemistry
Math. 340, Differential Equations	Chem. 343, Physical Chemistry Lab
Humanities Elective	Humanities Elective

Senior Year

Fall	Spring
Ch.E. 443, Design I	Ch.E. 444, Design II

Ch.E. 453, Thermodynamics
Ch.E. 441, Seminar
Ch.E. 333, Unit Operations Lab II
Ch.E. 325, Unit Operations III
E.M., Dynamics
Phys. 361, Atomic Physics
Ch.E. Elective
Ch.E. 463, Projects
E.M. 310, Mechanics of Materials
E.E. 390, Electrical Circuits and Machines
E.E. 396, Electrical Equipment Lab
Humanities Elective

Graduate courses in chemical engineering had been offered as early as the 1941–1942 school year, but the first earned M.S.Ch.E.s were not produced until 1954. The number of graduate-level courses in chemical engineering increased during the fifties as the number of students taking them grew. In the 1948–1949 school year, only two classes were offered on the graduate level: Ch.E. 503, Research Methods and Administration (later titled "Technical Administration"), taught by Colonel Barker and focusing on the "means and methods of planning, supervising, coordinating, and financing cooperative research and development activities"; and Ch.E. 506, Research, supervised by Barker or Dr. Oxford.

In 1950, Oxford began offering Petroleum Technology as a graduate course, and the department as a whole offered direction of graduate theses. By 1956, Heat Transfer, Wood Products Technology, and Fluid Flow had been added to the graduate course offerings. Reactor Design was offered for the first time in 1957, and Nuclear Reactor Laboratory and Applied Unit Operations were added in 1958.

The increases in graduate student enrollment in the chemical engineering program correlated to an increase in research. Departmental research fell under the supervision of a number of university administrative units during the 1950s; the oldest of these was the Engineering Experiment Station. The station had been created by the Board of Trustees in 1920 "to make investigations and study engineering problems of general interests to the people of Arkansas, to serve the mechanical industries of the state, and the urban population . . . and to solve engineering problems for the agricultural interests of the state." The station did not achieve those goals during its first two decades, however, and it fell by the wayside during World War II. In the war's aftermath, the rapid influx of engineering students, coupled with the scarcity of qualified teachers, kept the station from returning to its original research mission, but by 1949 the College of Engineering decided to make a concerted effort to revitalize the station.

That year, Dr. W. B. Stiles was named associate director of the Engineering Experiment Station, and the reactivated station began once again to coordinate research activities for the college. As the 1950s moved on, the station won contract work for the U.S. armed forces and private industries in addition to the research projects carried on for the good of the state. In 1952, for instance, the station began a joint project with the Office of Air Research to develop a means of instantaneously analyzing gases. Colonel Barker and Dr. Oxford both were employed part-time on this project.

The U of A chemical engineering faculty produced a large percentage of the Engineering Experiment Station's output during the early years of its reactivation. Assistant Professor Lenoir carried out a research project entitled "Thermal Conductivity of Gas Mixtures." Colonel Barker directed a number of projects, including "Heat Transfer through Clothing," "High Quality Building Board," "Paper and Cellulose from Sericea," and "Light Weight Building Tile." Lenoir's work resulted in the first bulletin published since the station's reactivation; Barker's work on building board followed soon thereafter.

The reactivation of the experiment station in 1949 resulted in part from a demand for practical research into specific engineering problems faced in the public and private sectors. But, the station and its research programs also had a deeper purpose—it played a role in the protection and prosperity of the American people. "Our population is increasing at the rate of 2,500,000 per year," wrote Dean George Branigan in 1952, "so that we shall have around 175,000,000 people in the United States by 1960. It becomes obvious that we cannot continue to spend for defense and raise the standard of living of our expanding population without an expanding economy. New resources and substitute raw materials must be discovered, marginal resources must be utilized economically, and new processes developed for better production and distribution." The Engineering Experiment Station played a role in achieving those goals.

The Institute of Science and Technology, created by the board of trustees in 1948, oversaw and coordinated all research activities in the College of Engineering for six years, taking over for Ordark and the Bureau of Research. The institute emphasized, at first, industrial research, and the Engineering Experiment Station fell under its purview.

Director W. W. Grigorieff actively involved the Institute of Science and Technology in all research projects at the U of A, including, after 1949, projects in the humanities and social sciences. Within a few years, the institute's staff numbered thirty-six, not including graduate students, a few of whom were in chemical engineering, and clerical workers. The tremendous growth, in terms of both staff size and influence, however, brought a backlash of sorts against the institute. Fears that the institute was not providing adequate educational experiences—that its major focus was in providing research for the benefit of industry, not students—became increasingly evident during the early 1950s.

The situation came to a head in 1953, when Grigorieff resigned to take a position with the Oak Ridge Institute in Tennessee. After a brief interlude, the dean of arts and sciences took control of the institute in January 1954. A year later, the Institute of Science and Technology was discontinued.

Yet the University of Arkansas still needed a means of coordinating campus research. That responsibility was taken over by University Provost Lewis H. Rohrbaugh and his assistant, Marvin T. Edmison. When Edmison left for a similar position with Oklahoma State University in 1955, research coordination was assigned to the dean of the Graduate School. The dean, with the aid of an assistant, oversaw most research projects at the university for the next couple of decades. The role of the dean as research coordinator reflected the growing significance of graduate education and the increasing emphasis placed on graduate research at the university.

By the late 1940s, "the Barn," the temporary wooden building constructed during World War II and used for lab and class work by the Department of Chemical Engineering, had become overcrowded and out of date. Indeed, enrollment for the entire College of Engineering had swelled following the war, and the lack of sufficient space proved acute for all engineering departments. For the chemical engineers, this problem was compounded by the fact that their expensive laboratory equipment was housed in a building that could easily go up in flames, taking all their work with it. Despite the lack of space and permanency, chemical engineering students and faculty made the best use of what they had, as noted in the *Arkansas Alumnus:* "Work of exceptionally high caliber is being done in this department, despite the fact that the department is housed in a wartime emergency building."

In 1949, construction began on a new wing of Engineering Hall that would become the province of chemical engineers for the next three and a half decades. By the fall of 1951, the chemical engineers had moved into their new quarters, the first such accommodations at the U of A designed specifically for the study of their field. Professor W. K. Lewis of the Massachusetts Institute of Technology spoke at the official dedication ceremony, which took place just prior to the convocation for the 1952 Engineers' Day.

The $125,000 annex attached on the east side of Engineering Hall and ran in a north-south direction, creating the L-shape the building still has. Its basement held new unit operations laboratories; above these, an expanded engineering library, research rooms, and more laboratories made up the second floor. The third floor contained a plant design room, faculty offices, and more research rooms. The old chemical engineering faculty offices were converted into classrooms.

Engineering Hall Annex. (*Arkansas Engineer,* November 1950, cover)

By the late 1950s, the Engineering College's space became cramped once again. Enrollments of more than nine hundred in 1956–1957 and eleven hundred in 1957–1958 pressed Engineering Hall, which was originally designed to house about four hundred students, to its limits. Some relief came in 1957, when the Fayetteville Chamber of Commerce donated the Oberman Building. This three-story building, renamed the West Avenue Annex, became the home of a limited number of engineering and drawing classes.

Extracurricular activities also evolved during this period. Engineers' Day underwent a series of changes between 1948 and 1960. Notably, in

1953, a political rally was instituted at which engineering seniors vied for the title of St. Patrick, and selected coeds presented skits and talent shows for the title of St. Patricia. The day after the rally, which usually took place a couple of weeks before Engineers' Day, engineering students voted for their candidates.

The actual celebrations for Engineers' Day became more elaborate. The coronation ceremony for St. Patrick and St. Patricia took place at a dinner on the eve of the big day, as did an annual beard-growing contest and the awarding of the Order of the Golden Chicken to a selected professor and the Order of the Golden Shovel to certain students. One professor who received the 1956 Golden Chicken award remembered: "There has always been some confusion surrounding that award. . . . In the eyes of the faculty, it was considered to be an award for outstanding teaching. In the eyes of some students, at least some of the time, it was an award which went to the biggest chicken on a given faculty in a given year. I have always considered it one of my favorite awards of all time, although I don't list it in my vitae because no one outside the hills of Arkansas has ever

Engineers' Day, 1950s. Ann Easley sang her message to the engineers as she was helped out by a number of her Kappa sisters. (*Razorback,* 1954)

heard of it." An assembly was held at Engineering Hall the morning of Engineers' Day, followed by a procession to the Union, where the Knights of St. Patrick were ceremonially anointed. That evening was given over to the Engineers Ball.

Of course, the engineer-agri rivalry continued. One former student remembered that the rivalry introduced him to the chemical engineering department's pivotal figure. "My first encounter with Col. Barker was in my freshman year. It turns out that a group of engineering students had stenciled green shamrocks on the sides of the Agri. Building—6 feet up and 6 feet apart, all the way around the building. Col. Barker put out the word that he wanted a sample of the paint—As a naïve freshman, I brought him a sample. Ultimately the shamrocks were removed by sanding—solvents must [have] made the paint run in the sandstone. If you go over to the Agri. Building today and examine the outside walls, you will see the location of the sanding on this sandstone building. In hindsight, I am surprised Col. Barker didn't take punitive action against the 'carrier of the sample.'"

Engineers' Day faced opposition, despite its role as the highlight of the social year for engineering students. By the late 1950s, some university administrators and faculty began voicing displeasure at the increasingly risqué skits performed at the preelection rallies by candidates for St. Patricia. Other than that, the new university-wide Gaebale festivities threatened to undermine Engineers' Day as one of the important social events for the spring semester.

The U of A chapter of the American Institute of Chemical Engineers continued to grow between 1948 and 1960. Even at the beginning of the period, active membership numbered fifty-two. As enrollment increased in the chemical engineering program, and as new faculty members arrived on campus and new facilities were built, the AIChE further increased its membership and activities.

Alpha Chi Sigma continued to play an important role in the lives of many chemical engineering students; meeting regularly for lunch or dinner at eateries such as A.Q. Chicken House or the Midway Café, the Alpha Sigma chapter listened to outside speakers and planned events and activities. One event, the laboratory banquet, was described in the *Arkansas Engineer:*

> Dr. Amis' physical chemistry lab could hardly recognize itself when

> scrubbed and all decked out with linen for the lab banquet. Beakers rinsed out with potassium cyanide (to kill all the germs you know) were used as cups, and on the tables, weird pieces of chemical apparatus were used to hold the coffee, cream and sugar. Fingers were much in use as drumsticks, wings and other delicious parts were operated on by the skillful chemists and chemical engineers. Everyone went home full of fried chicken and corn—the corn courtesy of Bill Seaton and Dr. John Lenoir, amateur comedians.

In 1952, the U of A Alpha Chi Sigma chapter won the Efficiency Trophy, awarded biannually to the best nonhouse chapter of Alpha Chi Sigma. In 1957, the society honored Professor of Chemistry Lyman Porter "for loyalty and interest he has shown in the organization since its beginning on the University of Arkansas campus." By the end of the 1950s, the chapter had begun discussing the possibility of having its own house.

Although chemical engineering students shared unique difficulties and interests with other engineers, they also took part in an active student life at the University of Arkansas during the 1950s.

Athletics continued to provide a focal point for U of A students during this period. Coach John Barnhill's football Razorbacks posted back-to-back 5–5 records in 1948 (the first year Razorback football games were played in Little Rock's Arkansas Memorial Stadium) and 1949, despite the play of All-American tailback Clyde Scott. Following these disappointing seasons, Barnhill handed the reigns of the football team over to Otis Douglas and focused instead on his role as athletic director. After two sub-par seasons under Douglas, Barnhill brought Bowden Wyatt in as head football coach. Wyatt's teams, led by All-American guard William "Bud" Brooks, showed glimpses of greatness—including an appearance in the Cotton Bowl—but Wyatt left Arkansas for his alma mater, Tennessee, in 1954. Jack Mitchell coached mostly average Razorback teams through 1957, when Barnhill replaced him with the now-legendary Frank Broyles. Broyles came from the University of Missouri, and, after a ho-hum 1958 season, led the Razorbacks to a three-way tie for the Southwest Conference championship and a Gator Bowl victory in 1959. This was only Broyles's first step in making Arkansas a national, not just regional, football powerhouse.

Of the other major intercollegiate sports, basketball had the most success between 1948 and 1960. Yet the Razorback roundballers did not achieve the heights that they had earlier under coach Glen Rose, even after he returned to the campus in the 1950s. He did, however, lead the team to a conference championship during the 1957–1958 season. Arkansas resumed baseball as an intercollegiate sport beginning in 1947, but the team at first did not participate in Southwest Conference league play due to the travel difficulties involved. Even after it joined conference competition, the Arkansas baseball team could not compete with the Texas schools due to a lack of practice time thanks to the longer winters of northwest Arkansas. Arkansas's track and field teams posted respectable records during the period; the cross-country team won six Southwest Conference championships during the 1950s. Razorback golfers won the conference championship in 1958.

The 1950 opening of the university's Fine Arts Center brought new cultural opportunities to U of A students. For the first time, the university had proper facilities for showcasing local, regional, and national talents. Musicals, comedies, gallery exhibits, and concerts became common parts of campus life.

The University of Arkansas Department of Chemical Engineering made phenomenal strides under the leadership of Colonel Maurice Barker. Indeed, it was Colonel Barker's strong will and iron hand that set the stage for the successful decades to come. When Barker stepped down as department head in 1959, he handed over the reigns to Professor Charles Oxford, who served for two years. In 1961, Philip Bocquet arrived at the U of A to become professor of chemical engineering and chair of the department. Under Bocquet's guidance, the program continued to grow and to develop, and the faculty members he brought to the department influenced the lives and learning of students into the twenty-first century.

Biographical Sketches, 1948–1961

Colonel Maurice E. Barker arrived at the University of Arkansas in the fall of 1948 as professor of chemical engineering and head of the

Department of Chemical Engineering. At that time, Barker was the only faculty member in the College of Engineering with a doctorate.

Barker was born May 20, 1894, on a farm near Humboldt, Tennessee. Upon completion of high school in Tennessee, he entered Valparaiso University in 1912; he graduated with a bachelor of science degree in 1915. From 1915 to 1917, he took part in a federal program that served as a precursor to the modern Peace Corps, teaching physics and biology at Laguna High School and serving as director of athletics for Laguna Province in the Philippines. On August 9, 1917, Barker entered the Coast Artillery Corps of the U.S. Army as a second lieutenant; within a year, he had been promoted to captain. Between 1918 and 1924, he served as a battery officer, battery commander, and artillery engineer, at the same time completing a master's degree in chemistry at the University of South Carolina. In 1924, he transferred to the Chemical Warfare Service.

Colonel Maurice E. Barker. (*Armed Forces Chemical Journal,* 1953)

In the Chemical Warfare Service, Barker served one year in the Chemical Warfare School and two years as technical director and safety engineer at the Edgewood Arsenal in Maryland. At the same time, he worked under Professor W. K. Lewis of the Massachusetts Institute of Technology on a thesis involving the development of a practical process for the manufacture of activated charcoal from coal. In 1930, MIT granted Barker his Sc.D. While at MIT, he was initiated into the Alpha Zeta Chapter of Alpha Chi Sigma.

In 1930, Barker became head of the Research Division of Edgewood Arsenal. A year later, he became instructor of physics, chemistry, and chemical engineering with the U.S. Army's Chemical Corps School; in 1935, he began a tour with the U.S. Army 2nd Division. From 1938 to 1942, Barker served as director of the Research and Engineering Division, Chemical Warfare Service. Under his watch, the division witnessed a hundred-million-dollar expansion of manufacturing facilities and developed new gases, chemical munitions, and manufacturing processes.

In 1942, Barker joined the Western Task Force Staff and took part in planning the invasion of North Africa. When the Fifth Army was organized at Oujda, Morocco, in January 1943, Barker joined up and subsequently served in the Tunisian and Italian Campaigns. In June 1944, he returned to the United States as commandant of the Chemical Corps School and commanding officer of the Chemical Corps Training Center. In August 1948, Barker retired from the U.S. Army with the rank of colonel. His next destination: the University of Arkansas.

For more than a decade, Barker led the Department of Chemical Engineering with a firm hand and a resolute will. "He was a large man, a man [who] was tireless in whatever he attempted," remembered Professor Charles Oxford. "I think he taught me and the students more than we realized at the time. He went at everything as if he were killing snakes. He didn't take his time with anything—teaching class, deer hunting, moose hunting, whatever—and he was not a man to stand on protocol." In his unmistakable style, Barker at one time or another taught many of the classes in the chemical engineering curriculum; his chief offerings included Organic Technology, Wood Products Technology, and Plant Design, but at various times he taught Unit Operations, Metals Corrosion, and Technical Administration, and oversaw Chemical Engineering Seminar and Projects. He also directed graduate students in the completion of master's theses.

Students remember Barker as a tough professor, but one who had their best interests at heart. His focus lay in not only equipping students for work in chemical engineering but also in providing them with a foundation for all types of ventures. "One thing I remember him telling us," recalled a student from those years. "'You get a degree in chemical engineering, you don't have to be a chemical engineer. You can go do anything you want to do, and you're equipped probably better to do it than ninety-five percent of the people.' Well, at the time, [I] didn't pay too much attention to him, but after being out of college for a few years, and getting into the kind of work that I did, I said 'Doc Barker was exactly right.'"

Colonel Barker also knew that the department's primary focus lay in preparing undergraduates for successful professional careers. "Colonel Barker made it clear," remembered one former student, "he was training chemical engineering students to enter the working world following successful completion of their undergraduate work."

Barker and his fellow faculty members emphasized quality of work above quantity, acquisition of correct study habits, and investigation over remembered knowledge. His personal research centered on heat transfer through fibrous materials, with the ultimate goal of improving clothing for cold-weather wear, and the development of products and processes for the economical use of scrub timber in the manufacture of building boards and dissolving pulp for the production of rayon. In connection with this work, Barker pioneered the use of *sericea lespedeza* as a source of cellulose. The results of his research: nineteen U.S. patents, six books or bulletins, and over one hundred technical and popular publications.

In June 1959, Barker became the first emeritus professor of chemical engineering at the University of Arkansas. Professor Charles W. Oxford replaced Barker as head of the department. Barker remained emeritus professor until his death on November 29, 1979.

Charles "Epi" Oxford graduated with his B.S.Ch.E. from the University of Arkansas in March 1944 with graduation honors, senior class honors, and departmental honors. "I actually finished my work in '43," he recalled, "but my diploma's dated '44 because they didn't give mid-term diplomas at that time." After a brief stint (1943–1944) as an instructor in physics, algebra, and trigonometry at the U of A, a tour with the U.S. Navy, and two years at the University of Minnesota, Oxford ventured to the University of Oklahoma, where he earned a Ph.D. in chemical engineering in 1952. In 1948, he returned to his undergraduate alma mater, where he became assistant professor of chemical engineering. Nine years later, Oxford became professor of chemical engineering.

When Colonel Barker stepped down as department head, Professor Oxford took over on an interim basis until the arrival of Professor Philip Bocquet in 1961. This early administrative post was a sign of things to come. In 1963, Oxford became associate dean of the College of Engineering, a post he held until being named administrative vice president of the university in 1968. In 1970 his title changed to executive vice president; five years later he became vice chancellor for academic affairs.

By 1983, Oxford had been named vice president for academic affairs for the entire University of Arkansas System. Twice during his administrative career, in 1973 and again in 1983, Oxford served as acting president of the university. He retired from his administrative post and the Department of Chemical Engineering during the 1986–1987 school year, when he became emeritus professor and vice president.

Professor Oxford is a registered professional engineer in Arkansas and Kansas. His industrial experience includes stints with Esso, Union Carbide, and Boeing. He is a member of Alpha Chi Sigma, Tau Beta Pi, Pi Mu Epsilon, Omicron Delta Kappa, Sigma Xi, the AIChE, and the American Society for Engineering Education (ASEE). His research interests have included desorption of hydrocarbons, gas flow at extremely low-pressure, nondestructive testing, and stabilization of fuels. He has received an honorary Doctor of Laws degree from the University of Arkansas and an honorary Doctor of Science degree from the University of Arkansas at Little Rock.

Charles Oxford. (*Arkansas Engineer,* November 1968, 7)

Over the years, Professor Oxford taught a remarkably varied number of chemical engineering courses. His classes included Chemical Engineering Fundamentals and Materials, Petroleum Processing, Fuel and Gas Analysis Laboratory, Industrial Stoichiometry, Unit Operations, Chemical Engineering Thermodynamics, Chemical Engineering Seminar, Petroleum Technology, Heat Transfer, and Fluid Flow.

Students remember Oxford as "a real asset to the department." In the words of one former undergraduate, "the students all thought the world of 'Epi.'" Another former student, who later returned to the U of A as a member of the chemical engineering faculty, recalled: "My office mate when I joined the department was Charles 'Epi' Oxford, who had been my thermo professor when I was an undergraduate. Not only was he one of the three best teachers I ever had, he was a great mentor! He was kind enough to take a look at my exams and gave me valuable counseling about the craft of teaching. He also was a scholar and a gentleman."

James Riley Couper arrived at the University of Arkansas as associate professor of chemical engineering in 1959. Couper received his B.S.Ch.E., M.S.Ch.E., and D.Sc. from Washington University. Before his arrival at the U of A, he worked for the Missouri Portland Cement Company and for Monsanto; his experience there included six and a half years in process design and economics and one year as production supervisor of five manufacturing departments. Couper put his industrial experience to good use in the classroom. "We, the students, were always amused during his early years at U of A because everything was referenced to 'Monsanto,'" remembered one former student.

James Couper. (*Arkansas Engineer,* November 1969, 14)

Couper was promoted to professor of chemical engineering in 1965 and served as administrative assistant in the U of A Office of Research Coordination from July to September of 1968. In 1969 he became chair of the Department of Chemical Engineering. He returned to full-time teaching in 1979 and became professor emeritus in 1989. Even after his retirement, Couper has remained a familiar face around the department; on five occasions since 1989, he has taught the department's design courses or Technical Administration. He still teaches technical administration through the U of A Distance Learning Program.

Professor Couper has served as a consultant to NL Industries, Alcoa, Amoco, Dow, Phillips, and Keystone Appraisal Company. At one time or another, Couper taught every undergraduate course in the curriculum with the exception of Kinetics. Although he concentrated on design and economics courses (Design I and II), he also enjoyed teaching technology courses such as Polymers and Technical Administration. Outside the confines of the U of A, he taught over eighty process economics and/or cost estimation courses for AIChE, McGraw-Hill, the Center for Professional Development, and the Seminar Development Corporation between 1977 and 1993.

Couper's research interests focus on viscoelastic properties of asphalt, recovery of tar from tar sands, thermal properties of emulsions and suspensions, and economic optimization. He supervised the work of forty master's and nine Ph.D. students, and he has authored or coauthored twenty articles in refereed journals, contributed sections to two encyclopedias, and presented thirty invited lectures or presentations. In 1986, he was principal author with Professor William Rader of the U of A Department of Industrial Engineering of a text entitled *Applied Finance and Economic Analysis for Scientists and Engineers,* and in 2001, he was principal author with O. Thomas Beasley (B.S.Ch.E., 1951) and Professor W. Roy Penney of a text entitled *The Chemical Process Industries Infrastructure: Function and Economics.* At present, Couper is working on a book entititled *Process Engineering Economics.*

Professor Couper is a member of the ACS, the AIChE, the American Association of Cost Engineers, Alpha Chi Sigma, Tau Beta Pi, and Sigma Xi. He was instrumental in establishing Tau Sigma, a local chemical engineering honorary fraternity that later became a chapter of Omega Chi Epsilon, national chemical engineering fraternity, with which he served as treasurer, vice president, and president. He is a registered professional engineer in Missouri and Arkansas, and, from 1988 to 2000, he served on the examination committee for the National Council of Examiners for Engineering and Surveying. Couper also is a member, and served for a decade as advisor, of the U of A chapter of Sigma Chi fraternity.

Professor Couper has served as chairman of the ASEE Department Heads Forum and was secretary, vice chairman, and chairman of the Industrial and Engineering Chemistry Division of the American Chemical Society. In August 1986, the AIChE elected him to Fellow grade. Among his many other honors have been inclusions in American Men and Women of Science and numerous "Who's Who" listings.

James Stice became assistant professor of chemical engineering in 1954. Almost five decades later, he recalled the fortuitous circumstances of his hiring:

> I was working for Thurston Chemical Company in Joplin, MO, a sort of family company which had recently been acquired by W. R. Grace and Co. Most of the officers of the company were incompetent, and many of their personnel policies were really bad. Today, many of those policies would have them in hot water with OSHA and the Wage and Hour boys big time. Anyhow, I had decided to look for another job elsewhere.
>
> Just about that time, one of the three profs at Arkansas quit. He was Tom Leland, a very bright guy indeed. As I recall, he had done his PhD work in low-temperature processes. . . . Now it happens that Rice University had a Japanese-American prof named Ricky Kobayashi, who was doing low-temperature work, and he had research money! In 1954, nobody had research money. Ricky also had a brain tumor, and they operated on him; folks thought he wouldn't survive, so they [Rice] wanted someone aboard to carry on his research. They offered Leland a job, and I guess it was one of those deals he couldn't refuse. So he went to Rice.
>
> Arkansas was in a bind—they had just lost 33 percent of the ChE faculty, and it was too late to mount a national search for a replacement. Thus, at around 6 a.m. some Tuesday, I got a phone call from Colonel Barker, who knew I wasn't happy with my job. He told me what had happened, offered me the job, and said that it was only for a year, until they could conduct a proper search. In the meantime, I could use the placement office to find another job, and I wouldn't have to sneak around doing it on weekends. What he didn't know was that my grandparents had both died in 1953, and their house at 124 West Prospect was standing vacant. We could live there for free, and bank more of my salary, which was within $100/yr of what I was making in industry. I told him I'd call him back, Patsy and I talked about it, and around noon I accepted.
>
> I never intended to be a college teacher. I was raised in Fayetteville, and I knew a lot of the professors' kids. So I knew something about the families. None of them was getting rich, and I had entered industry with

the hope of becoming rich and famous—in a pinch, I'd settle for rich. But I jumped into the job with enthusiasm.

I had been used to working 55–60 hours per week in industry. My first semester, Col. Barker gave me 12 hours to teach. My colleagues at Thurston had asked me what I'd be doing when I became a prof. I told them I'd have to teach 12 hours per week. They asked what I'd be doing with the rest of my time? I fatuously told them that I'd probably catch up on my fishing and golf! Those 12 hours turned out to be Stoichiometry, Unit Ops I, Unit Ops II, Fuel and Gas Lab (3 afternoons per week), and Unit Ops Lab (two afternoons per week)—that's 28 contact hours per week! Jesus. I never worked so hard in my life! My workload went up to around 80 hours per week, and for the first two or three months I thought I was going to die. I also discovered that I loved it.

Stice had received his B.S.Ch.E. at the University of Arkansas and, later, both his M.S.Ch.E. and Ph.D. at the Illinois Institute of Technology. He remained with the Department of Chemical Engineering until 1957,

James Stice

when he left for a five-year tenure as instructor in chemical engineering with the Illinois Institute of Technology. In 1962, Stice returned to the U of A as associate professor, then full professor. In 1968, he joined the Department of Chemical Engineering at the University of Texas, Austin.

Stice has received numerous awards and distinctions throughout his career, including a National Science Foundation Science Faculty Fellowship, the ISA Journal Award, the Chester F. Carlson Award for Innovation in Engineering Education, the Donald E. Marlowe Award for distinguished administrative leadership in engineering education, five awards as Outstanding Chemical Engineering Professor at Texas, and an award from the UT College of Engineering as Outstanding Departmental Advisor. In 1995 he was named a Distinguished Alumnus of the University of Arkansas.

Stice has published fifty-five books or articles and has presented 162 invited addresses before scholarly and professional audiences. He has also given 226 workshops for university faculty and industrial trainers.

Chapter Five

The Bocquet Years, 1961–1969

The 1960s were a time of evolution, and sometimes revolution, for colleges and universities in the United States. The University of Arkansas did not face as many of the difficulties as other campuses—student rebellions, large anti-Vietnam demonstrations, or integration riots, for example—but it did reflect some of the more substantial changes on the American higher-education landscape. During this period, graduate education came to be emphasized more and more over undergraduate at many American universities; the U of A also increased the number of graduate programs it offered during the sixties—the College of Engineering began offering courses leading to the Ph.D. During the 1960s, the volume of research conducted by university faculty and students tripled, while overall enrollment at the Fayetteville campus doubled.

Related to this was increased federal support for research and graduate education. Besides increases in federal funding, private donations to the University of Arkansas increased enormously over the course of the decade. Outside funding of scholarships, student loans, endowed chairs, construction, and graduate fellowships became much more common than ever before. An increasingly active Alumni Association took the lead in searching for such funding. These developments would be reflected in chemical engineering with the creation of the Alcoa Distinguished Professorship.

The most striking changes at the U of A during the 1960s came in the form of the physical environment. Student housing shortages coupled with increased enrollments created the impetus for new dormitories; increases in holdings pushed the university to build a new library; outdated buildings in education, the sciences, engineering, and other fields resulted in new classroom and laboratory buildings; a new Health Center opened for the use of students and faculty.

The "space race," which began during the 1950s but picked up steam during the 1960s, and the overall greater complexity of U.S. industry placed an increasing demand on colleges and universities for more and more engineers. These demands created apprehension within most engineering colleges, where enrollments actually had been declining. The U of A College of Engineering was no different. Dean George Branigan expressed his worry in March 1962: "In a day when technology is paramount in our industrial society and skills are mandatory to implement science and engineering, this decline in engineering enrollments is of very vital concern." Branigan pointed to a number of reasons for the smaller enrollments, including the difficulty of engineering curricula, the lack of adequate high school preparation, and the unglamorous, noncreative public image of engineering.

Despite these problems, engineering enrollment at the U of A rebounded by the mid-1960s. The increases in student numbers did not, however, solve all of the problems of the College of Engineering. By the early 1960s, the U of A faced a double difficulty in engineering faculty recruitment and retention: the college did not offer competitive salaries and lacked adequate, up-to-date classroom and laboratory space. These problems allowed industry and other universities to lure away some of the college's most respected faculty members. As U of A president David Mullins noted in February 1962, "the inability to employ Ph.D.s at existing salary levels presents handicaps that must be soon overcome."

Despite the fact that it did lose some faculty to other universities and to industry, the Department of Chemical Engineering made significant, and enduring, additions between 1960 and 1969. In fact, the arrival of professors such as Jim Turpin, Philip Bocquet, Robert Babcock, Charles Springer, and Louis Thibodeaux marked the creation of a stable, enduring core of teachers and researchers for the next few decades. In terms of training, the chemical engineering faculty compared very favorably with other departments in the College of Engineering: by 1970, when only 60 percent of the college's faculty held the doctorate, all Department of Chemical Engineering members held terminal degrees.

Late in the decade, new forms of outside revenue allowed the creation of additional professorships within the College of Engineering and the Department of Chemical Engineering. In 1968, Mrs. Irma F. Giffels donated one hundred thousand dollars to the college to fund the first endowed chair in engineering. Later that year, the Aluminum Company of America donated money to create the Alcoa Distinguished Professorship in Engineering, to be filled by a chemical engineer. Alcoa promised an eight-thousand-dollar grant for each of the next five years and kept open an option that the funding could become permanent some time in the future. Lloyd A. Alexander became the first Alcoa Distinguished Professor of Chemical Engineering, serving for the 1968–1969 school year.

The following list of faculty who arrived during the period 1961–1969 includes many names that students from the sixties, seventies, eighties, and nineties should recognize—a testament to the wise personnel decisions made under the chairmanship of Professor Bocquet.

1. Philip E. Bocquet. Professor and Head of the Chemical Engineering Department (1961–1969); Associate Dean of Engineering and Associate Director of Engineering Experiment Station (1969–1977); Emeritus Professor (1985–1992). B.S.Ch.E., Texas A&M; M.S.Ch.E., Ph.D., University of Michigan.

2. Jimmy Lee Turpin. Instructor in Chemical Engineering (1960–1965); Assistant Professor of Chemical Engineering (1965–1969); Associate Professor of Chemical Engineering (1969–1975); Professor of Chemical Engineering (1975–1995); University Professor (1995–). B.S.Ch.E., M.S.Ch.E., University of Arkansas; Ph.D., University of Oklahoma.

3. Robert E. Babcock. Assistant Professor of Chemical Engineering (1965–1968); Associate Professor of Chemical Engineering (1968–1974); Assistant for Research, Water Resources Research Center and Office of Research Coordinator (1969–1971); Assistant Dean for Research, Graduate School (1971–1975); Director, Water Resource Research Center (1971–1980); Professor of Chemical Engineering (1974–); Assistant Dean, College of Engineering (1980–1981); Head of Department of Chemical Engineering (1990–2001). B.S.Pet.E., M.S.Ch.E., Ph.D., University of Oklahoma; P.E.

4. Charles E. Springer. Assistant Professor of Chemical Engineering (1965–1969); Associate Professor of Chemical Engineering (1969–1976); Professor (1976–1993); Emeritus Professor (1993–). B.S.Ch.E., M.S.Ch.E., Ph.D., University of Iowa.

5. Louis J. Thibodeaux. Assistant Professor of Chemical Engineering (1967–1973); Associate Professor (1973–1979); Professor (1979–1985). B.S.P.-Ch.E., M.S.Ch.E., Ph.D., Louisiana State University.

6. Lloyd G. Alexander. Alcoa Distinguished Professor of Chemical Engineering (1968–1969). B.S.Ch.E., Ph.D, Purdue University.

7. Carl L. Griffis. Assistant Professor of Chemical Engineering (1968–1971). B.S.Ch.E., M.S.Ch.E., Ph.D., University of Arkansas.

The 1960s brought important changes to chemical engineering curricula in the United States. The 1961 publication of Bird, Stewart, and Lightfoot's book, *Transport Phenomena,* created a renewed emphasis on mathematics for chemical engineering students and largely displaced the tried-and-true method of unit operations in many universities and colleges. The Department of Chemical Engineering at the U of A, however, stayed the unit operations course.

Significant changes were made to the curriculum, though. In 1961, the College of Engineering began requiring all students to take eighteen elective hours in the humanities or social sciences. Also that year, the college decreed that engineering students other than those in mechanical and industrial engineering would no longer be required to take shop courses. Inside the classroom, professors began to experiment with new teaching methods, including using films, slides, and other visual aids, and guest lectures. Along with these changes came an increased emphasis on graduate student education, and graduate-level chemical engineering courses became more diverse and numerous. It was also during this period that the College of Engineering began offering coursework toward the Ph.D.

Chemical engineering students began the 1960s taking a course load very similar to those at the end of the 1950s. Overall, the curriculum for the 1960–1961 school year simply rearranged some of the yearly coursework. Two new courses entered the curriculum, however: Ch.E. 4033, Process and Kinetics became a required course for the last semester of the senior year, and Modern Physics replaced Atomic Physics. The chemical engineering curriculum for the 1961–1962 school year looked like this:

Freshman Year

Fall	**Spring**
English 1013, Composition	English 1023, Composition
Chem. 1104, General Chemistry	Chem. 1114, General Chemistry
G.E. 1000, Engineering Orientation	Math. 2555, Calc. I & Analytic Geometry
Math. 1284, College Algebra and Trigonometry	I.E. 1021, Engineering Methods

I.E. 1122, Engineering Graphics I
Humanities-Social Sciences Elective
Military or Air Science

Humanities-Social Sciences Elective
Military or Air Science

Sophomore Year

Fall

Phys. 2053, Engineering Physics
Phys. 2071, Laboratory Physics
Math. 2565, Calculus II and Analytic Geometry
Ch.E. 2011, Fuel and Gas Analysis Lab
Ch.E. 2622, Fundamentals
I.E. 1132, Engineering Graphics II
Humanities-Social Sciences Elective
Military or Air Science

Spring

Physics 2063, Engineering Physics
Phys. 2081, Laboratory Physics
Math. 2573, Calc. III & Diff. Equations
Chem. 2224, Quantitative Analysis
Ch.E. 2224, Stoich. and Elem. Thermo.
Humanities-Social Sciences Elective
Military or Air Science

Junior Year

Fall

Ch.E. 3233, Unit Operations I
Chem. 3604, Organic Chemistry
Chem. 3403, Physical Chemistry
Chem. 3501, Physical Chemistry Lab
E.M. 2003, Statics
Phys. 2613, Modern Physics

Spring

Ch.E. 3243, Unit Operations II
Ch.E. 3322, Unit Operations Lab I
Chem. 3614, Organic Chemistry
Chem. 3413, Physical Chemistry
Chem. 3511, Physical Chemistry Lab
Econ. 2013, Principles
Math. 3403, Differential Equations

Senior Year

Fall

Ch.E. 4433, Design I
Ch.E. 4533, Thermodynamics
Ch.E. 4411, Seminar
Ch.E. 3253, Unit Operations III
Ch.E. 3332, Unit Operations Lab II
E.M. 3003, Dynamics
Humanities-Social Sciences Elective

Spring

Ch.E. 4443, Design II
Ch.E. Elective
Ch.E. 4632, Projects
Ch.E. 4033, Process and Kinetics
E.E. 3903, Electrical Circuits and Machines
E.E. 3961, Electrical Equipment Lab
E.M. 3103, Mechanics of Materials

For the first time, freshman chemical engineers had the option of qualifying by examination for Mathematics 1284, College Algebra and Trigonometry. Upon passsing the exam, the freshman were allowed to skip that course and take Mathematics 2555, Calculus I and Analytic Geometry, during the fall semester, thus enabling them to finish their mathematics requirements a semester early. Students chose the senior-year Chemical Engineering Elective from one of the following: Ch.E. 4123, Materials of Construction; Ch.E. 5033, Technical Administration; Ch.E. 5053, Reactor Design and Operation; or Ch.E. 5223, Petroleum Processing.

The faculty made no changes to the curriculum during the 1961–1962 school year, but it added new electives—Ch.E. 488V, Special Problems; Ch.E. 5023, Advanced Transport and Processes; Ch.E. 5403, Organic Technology; and Ch.E. 5413, High Polymer Theory and Practice—the following year. Engineering Graphics II was dropped, and Ch.E. 2623, Fundamentals, was increased from two to three hours for 1963–1964. The following year, chemical engineering seniors had the option of taking another new elective, Ch.E. 4303, Automatic Process Control.

Major changes in the curriculum came for the 1965–1966 school year. Mathematics 2555, Calculus and Analytic Geometry, replaced Mathematics 1284, College Algebra and Trigonometry, as a first semester course. Also in 1965–1966, the Organic Chemistry and Statics courses moved to the sophomore year; Ch.E. 2011, Fuel and Gas Analysis Lab, was dropped completely; Dynamics and an extra chemical engineering elective were added and Physical Chemistry Lab dropped from the junior year; and Unit Operations Lab became Chemical Engineering Lab.

No changes were made to the curriculum for the 1966–1967 school year, but 1967–1968 saw a variety of adjustments. That year, nine hours of the humanities-social sciences electives became prescribed courses: Western Civilizations 1003 and 1013 and World Literature 1113. Thermodynamics moved to the junior year, replacing Modern Physics, which was dropped altogether from the curriculum, and three hours of chemical engineering electives were added to the senior year.

By 1969, the chemical engineering faculty had made further refinements to the curriculum. These changes included replacing General Chemistry with College Chemistry and Fundamentals of Chemical

Engineering in the freshman year, moving Differential Equations to the junior year, and adding a new course to the junior year—Ch.E. 4543, Physical and Chemical Equilibria. By the end of the Bocquet years, the chemical engineering curriculum at the University of Arkansas looked like this:

Freshman Year

Fall	Spring
English 1013, Composition	English 1023, Composition
Chem. 1125, College Chemistry	Math. 2565, Calc. and Analytic Geometry
Math. 2555, Calculus and Analytic Geometry	Western Civilization 1003
G.E. 1122, Engineering Graphics	Ch.E. 2123, Fundamentals
G.E. 1000, Engineering Orientation	

Sophomore Year

Fall	Spring
Phys. 2053, Engineering Physics I	Phys. 2063, Engineering Physics II
Phys. 2071, Laboratory Physics I	Phys. 2081, Laboratory Physics II
Math. 2573, Calculus and Differential Equations	Western Civilization 1013
Ch.E. 2133, Elementary Thermodynamics	Chem. 3614, Organic
Chem. 3604, Organic	Math. 3403, Differential Equations
Military or Aerospace	E.S. 2003, Statics
	Military or Aerospace

Junior Year

Fall	Spring
Ch.E. 3233, Unit Operations I	Ch.E. 3243, Unit Operations II
Ch.E. 3312, Chemical Engineering Laboratory I	Ch.E. 3322, Chem. Engr. Laboratory II
Chem. 3404, Physical	Chem. 3414, Physical
Ch.E. 4533, Thermodynamics	Econ. 2013, Principles
E.S. 3003, Dynamics	Ch.E. 4543, Phys. & Chemical Equilibria
World Literature 1113 or Philosophy 2003	

Senior Year

Fall	Spring
Ch.E. 4433, Design I	Ch.E. 4443, Design II
Ch.E. 4133, Chemical Processes and Kinetics	Technical Electives (6 hours)
Ch.E. 4411, Seminar	Ch.E. 4632, Projects
Ch.E. 3253, Unit Operations III	E.E. 3903, Electric Circuits and Machines
Ch.E. 3332, Chemical Engineering Laboratory III	E.E. 3961, Electrical Equipment Laboratory
E.S. 3103, Mechanics of Materials	Humanities-Social Sciences Elective
Humanities-Social Sciences Elective	

Graduate students became more common in the chemical engineering department between 1961 and 1969, and the Bocquet years brought the first engineering Ph.D.s granted by the University of Arkansas. George D. Combs earned the first Ph.D. in engineering in 1964, but it was a year later before Robin Moore earned a Ph.D. in engineering with an emphasis on chemical engineering. Earning a Ph.D. was not without its difficulties, Moore remembered: "Money for research was a big problem. I had completed all my requirements for the degree except the research, but could not proceed because we did not have the money to buy measurement equipment. After waiting nearly a year, [the] National Science Foundation, finally, came up with some money to buy equipment. I literally lost a year of time waiting for grant money."

During the late 1960s, many engineering students began to express concerns about the difficult nature of their educational programs. An editorial in the March 1969 issue of the *Arkansas Engineer* called for expanding to nine semesters degree plans for all engineering fields, noting that it usually took the average student five years to complete the program anyway. An added benefit, for students who worried about being drafted into the military, would be a student deferment extension if the college officially changed the programs from eight to nine semesters. "Students would find it possible to take a greater part in University activities as well as to devote more time to studies," noted the editorial, "without pressure from the Selective Service System. This would also afford more opportunity to take elective courses from other colleges, possibly alleviating the oft-cited complaint that engineers lack a well-rounded education."

Whatever a well-rounded education might entail, one development is evident in looking at the 1960s: The Department of Chemical Engineering continued to emphasize research as a method of instruction, a means of gaining recognition, and an opportunity for service to the state. With the opening of a new Science-Engineering Center in 1964, and the subsequent relocation of the departments of electrical and mechanical engineering to their new home, chemical engineering devoted more space to both faculty and graduate student research. Additionally, money for research increased tremendously throughout the 1960s for the university as a whole, and chemical engineering shared in the abundance.

A key element of the research program during this decade came from state-funded initiatives. In the mid-1960s the Arkansas State Highway Department and the Bureau of Public Roads funded a three-year project investigating "the flow properties of asphalt as related to its composition." In an effort to develop and maintain better roads in the state, the project, under the guidance of James Couper and funded with forty-three thousand dollars of state money, sought "methods for predicting the performance of asphaltic materials used on the roads of this state."

By 1967, two other major research projects—Couper's study of "Thermal Conductivity of Two-Phase Systems" and Bocquet's on "Measurement of Streaming Potentials," both funded by the National Science Foundation—raised the total outside funding for chemical engineering research projects to nearly $150,000.

All of the department's research came under the sponsorship of the Engineering Experiment Station. Under the leadership of its director, Dean of Engineering George Branigan, and its associate director, Associate Dean of Engineering Charles Oxford, the experiment station helped researchers prepare and submit proposals, find space and equipment, and write and present reports. The station also distributed bulletins to college and public libraries, individuals, and industry. By the mid-1960s, the Engineering Experiment Station coordinated research activities funded at a half million dollars and provided researchers with a shop staffed with trained machinists. "Without research and publication," noted Oxford in an *Arkansas Engineer* article, "a university is not worthy of its name; without research, graduate students could not be adequately trained. The Engineering Experiment Station exists to foster these activities in an orderly fashion."

The most significant change in the College of Engineering's physical environment during the sixties only tangentially impacted the Department of Chemical Engineering. In 1964, a new Science-Engineering Center opened to the west of Engineering Hall and south of the Chemistry Building. The new $2.5 million center housed classrooms, laboratories, and faculty offices and became the electrical engineering and mechanical engineering departmental headquarters. It also gave new space to the graphics section of industrial engineering.

The Science-Engineering Center actually consisted of three separate buildings: a 100,000-square-foot, seven-story building housing the departments of electrical engineering, mathematics, and zoology, plus the Computing Center; a 30,000-square-foot Mechanical Engineering building; and an instruction auditorium with a capacity of more than five hundred. The Computing Center, which had housed an IBM 650 in the Physical Sciences Building since 1960, received a new state-of-the-art IBM 7040 to complement its new home. The Science-Engineering Center was supplemented by the Science "D" building, which opened in the late 1960s.

Although the Science-Engineering Building directly benefited the departments of electrical engineering and mechanical engineering, students and faculty in civil engineering, industrial engineering, engineering mechanics, and chemical engineering received new advantages as well. The new building allowed greater space in Engineering Hall for the increasing enrollments most engineering departments faced after the mid-sixties. More space also correlated to an increase in faculty and graduate student research, which in turn opened the door for more outside funding for research projects. The extra research facilities also helped in recruiting faculty to the university's engineering departments.

Despite the demands of a rigorous course load and, increasingly, research projects, student engineers at the U of A made time for extracurricular activities. The Engineers' Week rallies came under more and more fire for the use of off-color entertainment. By 1965, calls for an end to the rally were being addressed in an *Arkansas Engineer* article, which stated that it would be necessary to tone down the risqué routines or to face the inevitable demise of the tradition. Later in the same issue, Engineering Dean George Branigan threatened to eliminate the rally if better judgment were not shown. On top of these concerns, the same article called for better planning of all of the week's events. Although some of these issues were addressed, by the end of the decade the turnout of spectators and participants for Engineers' Week was becoming increasingly smaller.

Despite these problems, Engineers' Week continued to be an important part of the school year for many students, and it continued to evolve. In 1964, Bobby Jones became the first chemical engineering student in decades to be elected St. Patrick. Inductions into the College of Engineering Hall of Fame began in 1965. Two years later, in 1967, it was decided that the use of green paint to decorate campus with clovers had to be done away with—the damages simply became too costly—and that another way of proclaiming Engineers' Week should be developed; one of the possibilities discussed included building a double-arch bridge connecting Old Main to the library.

Participation in traditional extracurricular activities such as Engineers' Week did not stop students from creating new ones. In 1964, for example, chemical engineering students at the University of Arkansas founded Tau Sigma, a local chemical engineering honorary society dedicated to recognizing scholastic achievement, leadership, and character. Charter members included Robert Jones, Ronald Glass, Ronald Embry, Terry Taylor, Richard Remke, Ray Owen, Douglas Proctor, Lynn Leck, and Thomas Sanders. Professor James Couper served as its first faculty advisor. Over the next couple of years, Tau Sigma's activities expanded into areas designed to help novice chemical engineers, including tutorial services for freshmen and sophomores, and, by January 1967, the society decided it had grown enough to seek a charter from Omega Chi Epsilon, the national chemical engineering honorary society. By March 1968, members could note that "as it looks, Tau Sigma will probably affiliate [with Omega Chi Epsilon] sometime next year."

Older societies continued to be active, too. For the first time ever, the Arkansas chapter of the American Institute of Chemical Engineers sent representatives to the national convention in 1964. The following year the chapter held its first annual Christmas party; the celebration included humorous skits and presents for the chemical engineering seniors. In 1966, the chapter for the first time won the Open House competition during Engineers' Week and hosted the regional AIChE meeting.

Outside of extracurricular activities specific to their field, chemical engineering students witnessed and took part in an increasingly diverse campus life. The 1960s are most often recalled as a period of turbulence and unrest in the United States as a whole and particularly on college and university campuses. Although the University of Arkansas was not immune to the disruptions that marked many other campuses during the decade, the atmosphere in Fayetteville remained relatively calm. Only occasionally did students demonstrate or protest in support of the free speech, peace, or civil rights movements. As one former student remembered, the chemical engineers did not spend much time worrying themselves with such displays: "As so often is the case, I think you will find that only a very, very few *engineers* were embroiled in these activities."

The greatest show of student discontent at the U of A came during the mid-sixties as a result of various efforts by the school's administration to monitor the process by which outside speakers were invited onto the campus. Students protested in 1966 when the school's printing shop refused to print *Preview,* a university literary magazine, because of scatological and homosexual references in the poems of one of its contributors. Like students at hundreds of other colleges and universities in the United States, students protested against the Vietnam War, most noticeably by "picketing for peace" during the noon hour. In 1968, the Southern Student Organization Committee engaged in its first protest at the university when it picketed against recruiters from the Dow Chemical Company. Finally, the first organized black protests on the University of Arkansas campus occurred in the wake of Martin Luther King Jr.'s April 1968 assassination. Days later, black students formed Black Americans for Democracy (BAD). On May 14, BAD led a protest against the university's student newspaper, the *Traveler,* blocking the entrance to Hill Hall, where the paper's offices were located.

The student protests and demonstrations of the 1960s reflected a changed sense of empowerment among students. Another reflection of changes happening at the University of Arkansas and in the United States in general was the decision to drop compulsory military drill for all male students, a requirement that had been part of the university's mission since its inception in 1871. In May 1968, a student-faculty committee reported that thirty-eight land-grant schools had gone to voluntary drill and that the U of A was one of only sixteen still forcing male students to participate. The committee recommended—and after lengthy debate, the Faculty Senate, University President David Mullins, and the board of trustees approved—a change to voluntary military drill on June 5, 1969.

The 1960s were the heyday of Razorback football. Aside from the rebuilding years of 1963 and 1967, Head Coach Frank Broyles led the team to unparalleled heights, including regular top-ten rankings, winning records, conference championship, and New Year's Day bowl games. The highlight of the decade came during the 1964 season, when the Razorbacks went undefeated for the regular season and beat Nebraska in the Cotton Bowl; most postseason polls named the team national champions. The following year, the Razorbacks once again went undefeated in the regular season but lost to Louisiana State in the Cotton Bowl.

The Razorbacks did not achieve such spectacular heights in other sports during the 1960s. The basketball team proved to be mediocre during this period, although standout Tommy Boyer developed into a legend between 1961 and 1963, when he became the top scorer in Arkansas history and set the national record for best free throw percentage. Razorback baseball faced the same old problems as it had earlier: distance from the other schools in the Southwest Conference and a shortened training season. Track during the 1960s produced respectable results, including a cross-country conference championship in 1966. Razorback golfer R. H. Sikes won the 1963 NCAA tournament, and, for the first time, the university fielded a swim team in 1966.

By the sixties, Fayetteville's population eclipsed twenty thousand. New apartment complexes and shopping centers sprouted up around town, as did new residential areas like Oakland Hills and Oaks Manor subdivisions to the north of town. A relatively new problem, lack of parking, became quite evident, especially on the square and on Dickson Street. Plans were made during this period for an industrial park south of town near Fifteenth and Armstrong streets. The city's population and economic growth would continue into the seventies.

When Philip Bocquet stepped down as department head in 1969, he left the position knowing that chemical engineering had blossomed under his care. The addition of faculty members who thrived on teaching and excelled at research, and who exhibited a long-term commitment to the department and to the University of Arkansas, promised a bright future. Bocquet ventured to Madrid, Spain, where he served as a Fulbright

Lecturer for the 1969–1970 school year, secure in the knowledge that James Couper now manned the departmental ship. Couper would preside over a ten-year period of remarkable stability.

Biographical Sketches, 1961–1969

Philip E. Bocquet arrived at the University of Arkansas in 1961 as professor and head of the chemical engineering department. He took control of a department that, despite rapidly increasing enrollments over the previous decade, consisted of only four full-time faculty members and a single secretary. By the time he stepped down in late 1968, Bocquet had overseen a doubling of the department's faculty size and a significant increase in staff support, and he had proved instrumental to the creation of the Alcoa Distinguished Professorship. To use Professor Bocquet's favorite adjective, his "uncanny" leadership set the stage for the department's future success.

Bocquet earned his B.S.Ch.E. from Texas A&M in 1940. He then ventured north to the University of Michigan, where he earned his M.S.Ch.E. in 1947 and his Ph.D. in 1953. Prior to his arrival at Arkansas, he had been associate professor of chemical engineering at the University of New Mexico in Albuquerque. In 1969, Bocquet served as a Fulbright Lecturer at the Universidad de Madrid, Spain. Upon his return to Arkansas, he replaced Charles Oxford as associate dean of the College of Engineering and director of the Engineering Experiment Station. He returned to full-time teaching in 1977 and retired in 1985.

Philip Bocquet. (ChemE collection)

Professor Bocquet was a registered professional engineer and a member of the AIChE, the ASEE, the ACS, and the American Association for the Avancement of Science (AAAS). He authored numerous journal articles and two monographs/translations of German works by Helmholtz and Smoulchowski, and he held three international patents for natural gas processes. He was a member of Sigma Xi, Phi Kappa Phi, and Tau Beta Pi. "Phil was a good teacher," remembered one of his colleagues, "He was interested in mechanical activities and computers. In fact, he developed some of the first computers that the department had . . . taught the kids how to use both analog and digital."

Jimmy Lee Turpin became assistant professor of chemical engineering at the University of Arkansas in 1966. Turpin earned his B.S.Ch.E. in 1960 and his M.S.Ch.E. in 1961 from the U of A and received his Ph.D. from the University of Oklahoma in 1966. While working on his bachelor's and master's degrees, he served as a graduate teaching assistant and an instructor in the Department of Chemical Engineering.

Turpin achieved the rank of associate professor in 1969 and professor in 1975. In 1995, his years of outstanding teaching were rewarded when he was named University Professor. Besides his academic work, Turpin often worked for such companies as Amoco, Monsanto, and Reynolds while on sabbatical or summer breaks.

At one time or another, Turpin has taught every course in the undergraduate catalog, with the exceptions of Reactor Design and Safety. He is a member of the AIChE, the ASEE (for which he has served as a campus activity coordinator, director, moderator, and panelist), Tau Beta Pi, Sigma Xi, and Pi Mu Epsilon. He has received numerous honors, including being named AIChE Student Chapter Outstanding Teacher fifteen times and winning the Halliburton Award for Excellence in Teaching seven times. Turpin's other honors include the American Gas Association Fellowship; the Arkansas Razorback Outstanding Faculty Award; the ASEE Outstanding Campus Activity Coordinator; the ASEE Young Faculty Representative, National Meeting; the Burlington Northern Award for Teaching; the Carnegie Foundation Arkansas Professor of the Year; the Catalyst Award for Excellence in Teaching, Chemical Manufacturer's Association; the Phillips Award for Outstanding Faculty; the Innovative Teaching Development Award, College of Engineering; the Texas

Instruments Outstanding Student Service Award; the University of Arkansas Alumni Association Outstanding Teacher Award; the University of Arkansas Teaching Academy, Founding Member; the DuPont Summer Grant for Chemical Engineering Teachers; the Ford Foundation Fellowship; the Jersey Production Research Fellowship; and the L. B. Smith Scholarship in Engineering.

Jim Turpin. (ChemE collection)

Professor Turpin has authored or coauthored seven articles in refereed journals, seven reports on proceedings, and six research reports. He has presented his findings at eight scholarly meetings and ten invited symposiums. Turpin also has served on over forty departmental, college, and university committees, while providing consulting work for many industrial projects and engaging in his own sponsored research projects. His contacts outside of academia allow him to open professional doors for students; he "spent a lot of hours helping seniors get jobs," remembered one graduate.

Perhaps one former student summed Turpin's service to the university and the Department of Chemical Engineering best: "It's hard for me to imagine, if somebody [were] to ask me, what could Dr. Turpin do that he'd be a better instructor, that he'd be a better example of a university professor, than what he already does, I'd be stumped. I couldn't even begin to say."

Robert E. Babcock became assistant professor of chemical engineering in 1965. He attained the rank of associate professor in 1968 and professor in 1974. Over the years, Babcock has served the University of Arkansas as assistant for research with the Water Resources Research Center and Office of Research Coordination (1969–1971); assistant dean for research of the Graduate School (1971–1975); director of the Water Resources Research Center (1971–1980); and assistant dean of the College of

Engineering (1980–1981). He served as head of the Department of Chemical Engineering from 1990 to 2001.

Babcock received his B.S. in petroleum engineering in 1959, his M.S.Ch.E. in 1962, and his Ph.D. in 1964, all from the University of Oklahoma. Before arriving at Arkansas, he served as senior research engineer with the Hydraulic Fracturing Project of the Esso Production Research Company in Houston, Texas.

Babcock has taught Thermodynamics I, Thermodynamics II, Advanced Thermodynamics, Mass Transfer I, Heat Transfer, Petroleum Engineering, and Fluid Mechanics.

Professor Babcock has been awarded five patents and has performed consulting work for a variety of companies. He is a registered professional engineer and has served as the president and secretary of the Arkansas Section of the American Water Resources Association. He also has been chairman of the directors of the Southern Plains River Basins Region and has chaired the Executive Committee of the Integrated Petroleum Environmental Consortium. He is a member of the American Society for Engineering Educators and the Society of Petroleum Engineers. Babcock has been named "Conservationist of the Year—Water" by the Arkansas Wildlife Federation and elected a Fellow of the American Institute of Chemical Engineers.

Robert Babcock. (ChemE collection)

Babcock has authored or jointly authored twenty-six publications and has made twenty-three presentations at professional meetings. He has been a member of twelve university committees and a representative to the Campus Council and the Council on Water Resources. He has also taught continuing education courses and provided national and state services for such groups as the

Mid-Continent Water Resources Research Directors and the Arkansas Stream Preservation Committee.

Charles Springer became assistant professor of chemical engineering in 1965, associate professor in 1969, and professor in 1976. He received his B.S.Ch.E., M.S.Ch.E., and Ph.D. from the University of Iowa.

Over his twenty-three years at the University of Arkansas, Springer taught numerous undergraduate and graduate courses, including classes on heat transfer, corrosion, and safety, plus various laboratories. His major area of research was the mitigation of environmental hazards. Among his published works were *Safety, Health, and Loss Prevention in the Chemical Process Industries* (coauthored with J. Reed Welker, with whom Springer delivered over fifty National Science Foundation-sponsored short courses on teaching process safety).

Charles Springer. (ChemE collection)

A colleague who worked with Springer for fifteen years remembered: "I think Charlie Springer, of all of us, was the best engineer. . . . You could go to him with a problem, and Charlie could figure it out. I think he was most probably the best engineer on getting things done and how to make things work that we had on the faculty."

Springer was famous for the coffee he made for the department. Invariably prepared by the time most other faculty members arrived at the office, it has been referred to as "the most invigorating coffee ever brewed." Springer retired in 1993 to "become a gentleman farmer in Hogeye."

Louis J. Thibodeaux was named assistant professor of chemical engineering in 1967. He had earned his B.S.P-Ch.E., M.S.Ch.E., and his Ph.D. from Louisiana State University. Thibodeaux became associate professor in 1972 and professor in 1977, and he remained at the U of A until 1984, when he returned to his alma mater as professor of chemical engineering.

As a graduate student, Thibodeaux received a graduate fellowship with the National Council for Air and Stream Improvement. Combined with increasing societal and professional interest in environmental matters, the fellowship sparked his interest in environmental engineering. At the U of A, that interest led to research and teaching on environmental aspects of chemical engineering. Thibodeaux became one of the earliest chemical engineering academics to actively search for an answer to the problem of reconciling chemical production with its inherent dangers. Close ties with the civil engineering department fostered this interest, and he conceived the idea of environmental chemodynamics while on sabbatical from the U of A during the fall of 1974.

Louis Thibodeaux. (ChemE collection)

Thibodeaux believed that he could not have done his pioneering work had he not been at the University of Arkansas: "If you go to those universities that consider themselves mainstream—and I have to admit University of Arkansas back then was not mainstream—to get promoted, you have to do what the other guys did, go down the same path . . . collecting a record, working on environmental things, and the types of publications I produced, which were not in . . . mainline chemical engineering journals . . . [there was no way] that I would ever have gotten promoted [elsewhere]." He added, "I have such a warm spot in my heart [for the U of A chemical engineering department] because of the things that I was allowed to do."

At the U of A, Dr. Thibodeaux taught undergraduates courses on

Chemical Processes and Kinetics and Unit Operations. On the graduate level, he guided students through Advanced Mathematics for Chemical Engineers, Transport Phenomena, and Advanced Reactor Design. He found time to publish his first book during his tenure at the U of A, also; *Chemodynamics: Environmental Movement of Chemicals in Air, Water, and Soil* came out in 1979 and is considered a standard work on the subject. A second edition was published in 1996.

While at the U of A, Dr. Thibodeaux authored or coauthored four chapters in four books, sixteen articles in scholarly or professional journals, and twenty-eight contributions to proceedings, reports, or bulletins. He presented his findings at fifty-eight professional meetings between 1967 and 1984. In 1982, he received the Distinguished Award in Teaching and Research from the Arkansas Alumni Association, the Halliburton Award of Excellence, and was named Outstanding Researcher in Chemical Engineering. He brought in outside research funding of more than a million dollars while at Arkansas, and he directed the work of thirty master's and three doctoral students.

Outside the classroom, Thibodeaux has served as a consultant and expert witness for private companies and public entities. A list of his committee work and university services would be too long to recount here. His professional memberships include the AIChE, American Water Resources Association, and Sigma Xi. He is a professional engineer in Louisiana and Arkansas.

Thibodeaux left the U of A in 1984 to become director of the U.S. Environmental Protection Agency (EPA) Center of Excellence in Hazardous Waste Research and professor of chemical engineering at LSU; he left that position to become director of the U.S. EPA Hazardous Substance Research Center, South/Southwest from 1991 until 1995. In 1990, he was named Jesse Coates Professor in Chemical Engineering at LSU.

Chapter Six

The Couper Years, 1969–1979

James Couper took over as head of the Department of Chemical Engineering in 1969 and served in that capacity over the next ten years, a period of history remembered as the decade of Watergate, inflation, and "malaise." Although the University of Arkansas and the chemical engineering department mirrored the national setting, they also continued to evolve and to progress toward their goals, both new and old.

In 1971, the University of Arkansas became part of a University System designed to bring about greater cooperation and better planning between the Fayetteville university and other state colleges. Included in the University of Arkansas System were: the University of Arkansas, Fayetteville; the University of Arkansas Medical Center; the University of Arkansas, Little Rock; the University of Arkansas, Monticello; the University of Arkansas, Pine Bluff; the Graduate Institute of Technology; the Industrial Research and Extension Center; the Graduate School of Social Work; the Little Rock Division of the School of Law; and five Graduate Centers.

Overall enrollment increases at the University of Arkansas slowed down during the early 1970s, but they remained significant enough to compound other problems. In 1977, University President Charles Bishop openly pondered how best to "achieve and maintain quality during a period of runaway inflation, burgeoning enrollments and increased operating costs?" Part of the answer was to increase student fees in 1976 for the first time since the beginning of the decade.

Enrollments in the College of Engineering dropped briefly at the beginning of the seventies, but they rebounded significantly by the end. In 1971, reflecting a constriction of the engineering job market, the number of freshmen enrolled in engineering courses declined 40 percent; the total number of engineering enrollees dropped 17 percent. Besides the tighter job market, the creation of engineering programs at other Arkansas institutions bit into enrollments. By 1972, however, predictions were being made of a shortage of qualified engineers by the end of the decade.

These predictions seemed to be coming true as early as 1975, when the *Arkansas Engineer* noted that the Engineering College's enrollment had increased 19 percent over the previous year and that graduating engineering students met with more and more job offers and higher starting salaries, an especially important consideration in a time of skyrocketing inflation. The enrollment increases were not only among undergraduates, either. By 1976, more than one hundred graduate students had entered the College of Engineering. In general, the increases in enrollment led to crowded classrooms and heavy teaching loads for professors.

Changes in the College of Engineering did not only come in the form of fluctuating enrollments, however. Gradually more and more students enrolled in engineering who, in previous decades, would have enrolled in nonengineering fields. Especially significant were the increasing number of women and African Americans enrolling in engineering courses. By 1975, women accounted for 6.6 percent and blacks for 2.0 percent of all engineering students. In addition to women and black students, foreign nationals also became more prominent. These trends mirrored national engineering education developments.

For chemical engineers, the seventies brought about an increasing search for relevance, in terms of both education and profession. Notably, some chemical engineers began to question their role in society and how chemical engineering might positively impact their surroundings. A new interest in the environment had blossomed during the 1960s, and, by the 1970s, environmental concerns had become a focus for much of mainstream America. The creation of the Environmental Protection Agency in 1970, and of the Office of Technological Assessment in 1972, placed environmental issues under the federal government's watch. As the 1970s wore on, chemical engineering curricula around the nation, and at the U of A, reflected this emphasis.

In the summer of 1975, the Engineering College created the Engineering Extension Center (EEC), yet another means to make its work relevant and more accessible to outsiders. The EEC coordinated all continuing education for state and local engineers. Professor Vernon McBryde directed the center's work from its headquarters in a converted apartment complex just across Dickson Street from Engineering Hall. The EEC offered seminars, workshops, conferences, and academic courses statewide, all in an effort to "disseminate new developments in engineering to professional engineers and others."

The Couper years were remarkable for the lack of turnover among the chemical engineering faculty. The foundations that had been laid under the chairmanship of Dr. Bocquet, and, even earlier, under Colonel Barker, remained steady throughout the 1970s. Between 1969 and 1979, only one faculty member, Carl Griffis, left the department, and he only crossed the hall to biological and agricultural engineering.

Besides the minimal loss of faculty, three new members were added early in the Couper years that would help define the department and its activities throughout the 1970s and beyond. Those three are

1. Charles M. Thatcher. Alcoa Distinguished Professor of Chemical Engineering (1970–1992); Interim Department Head (1989–1990), Distinguished Professor Emeritus (1992–). B.S.E., M.S.E., Ph.D., University of Michigan.

2. Jerry Havens. Assistant Professor of Chemical Engineering (1970–1975); Associate Professor (1975–1980); Professor (1980–1987); Distinguished Professor of Chemical Engineering (1987–). B.S.Ch.E., University of Arkansas; M.S.Ch.E., University of Colorado; Ph.D., University of Oklahoma.

3. Robert N. MacCallum. Assistant Professor of Chemical Engineering (1970–1975); Associate Professor (1975–1981). B.S.Ch.E., M.S.Ch.E., University of Texas; Ph.D., Rice University.

For the U of A as a whole, the 1970s brought a renewed focus on its "primary role of producing top-ranking undergraduates" while at the same time continuing to develop graduate programs—stressing quality over quantity in both cases. The Department of Chemical Engineering was no different. But the department also felt pressures from beyond the borders of the Fayetteville campus. Most important, the 1970s brought to fruition within the profession a concern with environmental and social issues that had been slowly developing during the previous decades. These influences can be seen in the development of more diverse and relevant offerings in terms of undergraduate technical electives and graduate-level courses focused on the environmental and societal impact of chemical engineering.

The first year of James Couper's chairmanship brought a few minor changes to the chemical engineering curriculum. That year, University Physics replaced Engineering Physics, and Electrical Equipment Laboratory was dropped.

Chemical Engineering Curriculum, 1969–1970

Freshman Year

Fall	Spring
English 1013, Composition	English 1023, Composition
Chem. 1125, College Chemistry	Math. 2565, Calc. and Analytic Geometry
Math. 2555, Calculus and Analytic Geometry	Western Civilizations 1003
G.E. 1122, Engineering Graphics I	Ch.E. 2123, Fundamentals
G.E. 1000, Engineering Orientations	Elective

Sophomore Year

Fall	Spring
Phys. 2053, University Physics I	Phys. 2063, University Physics II
Phys. 2071, Univ. Physics Lab I	Phys. 2081, Univ. Physics Lab II
Math. 2573, Calculus and Differential Equations	Western Civilization 1013
Ch.E. 2133, Elementary Thermodynamics	Chem. 3614, Organic
Chem. 3604, Organic	Math. 3403, Differential Equations
Elective	E.S. 2003, Statics

Junior Year

Fall	Spring
Ch.E. 3233, Unit Operations I	Ch.E. 3243, Unit Operations II
Ch.E. 3312, Chemical Engineering Laboratory I	Ch.E. 3322, Chem. Engr. Laboratory II
Chem. 3504, Physical	Chem. 3514, Physical
Ch.E. 4533, Thermodynamics	Econ. 2013, Principles
E.S. 3003, Dynamics	Ch.E. 4543, Phys. and Chemical Equilibria
World Literature 1113 or Philosophy 2003	

Senior Year

Fall	Spring
Ch.E. 4433, Design I	Ch.E. 4433, Design II
Ch.E. 4133, Chemical Processes and Kinetics	Technical Elective (6 hours)
Ch.E. 4411, Seminar	Ch.E. 4632, Projects

Ch.E. 3253, Unit Operations III
E.S. 3103, Mechanics of Materials
Humanities-Social Sciences Elective

E.E. 3903, Electric Circuits and Machines
Humanities-Social Sciences Elective

The department did not offer any new electives during the 1969–1970 school year.

Reflecting the continuity evident in other realms of the chemical engineering program, the course curriculum changed very little over the 1970s. For the most part, the undergraduate course of study followed the basic plan as outlined above, although some minor changes did take place. During the 1974–1975 school year, for example, superficial changes were made in the scheduling of courses—the one-hour Chemical Engineering seminar was moved from the senior year to the junior year and the Western Civilization and World Literature/Philosophy requirements became general humanities-social studies electives. The following year, Ch.E. 2123, Fundamentals, became Chemical Engineering I and Ch.E. 2133, Elementary Thermodynamics, became Chemical Engineering II.

Despite the minimal modifications to the undergraduate curriculum, changes did occur, especially in terms of courses offered as technical electives or graduate credit. Reflecting the state of chemical engineering as a profession and the needs of society in general and the chemical industries specifically, innovative courses were added and older courses were revised. For the 1971–1972 school year, Louis Thibodeaux began offering Ch.E. 4253, Rate Processes in Environmental Engineering, and Ch.E. 5133, Advanced Reactor Design. Two years later, Robert MacCallum began teaching Ch.E. 4803, Special Topics in Bioengineering, and the department dropped Ch.E. 5102, Nuclear Reactor Laboratory. The 1973–1974 school year saw the addition of Ch.E. 5753, Air Pollution, and Ch.E. 5801 and 6801, graduate seminars in which master's and doctoral students gave oral presentations "on a variety of chemical engineering subjects with special emphasis on new developments." A year later, Dr. Thibodeaux began offering Ch.E. 4263, Analytical Ecology Models. For the 1979–1980 school year, Ch.E. 5153, Advanced Chemical Engineering Calculations I and Ch.E. 5163, Advanced Chemical Engineering Calculations II were combined to form one course, Ch.E. 5153, Advanced Chemical Engineering Calculation.

Professor Jerry Havens played a significant role in the evolution of upper-level chemical engineering coursework by offering revised versions of older courses. Under his hand, Ch.E. 5253, Thermostatics and Thermodynamics, became Advanced Thermodynamics, focusing on "methods of statistical thermodynamics, the correlation of classical and statistical thermodynamics, and theory of thermodynamics of continuous systems"; Ch.E. 5303, Process Dynamics, became Advanced Automatic Control Theory, in which students considered "complex control systems, including cascade, optimal, direct digital control and adaptive control systems . . . with chemical engineering applications"; Ch.E. 4533, Chemical Engineering Thermodynamics, became Thermodynamics of Single Component Systems, in which "a detailed study is made of the thermodynamic 'state principle,' energy and entropy balances and their application to the solution of problems involving single component physical systems and processes"; and Ch.E. 4543, Physical and Chemical Equilibrium, became Thermodynamics of Multi-Component Systems, which extended the principles of Ch.E. 4533 "to allow treatment of multicomponent thermodynamics systems and processes" and considered "physical and chemical equilibrium in detail." Havens also offered a new course beginning in 1976, Ch.E. 4713, Energy Utilization Technology, "a survey of the energy supply and demand, past, present, and projected."

When Dr. Couper stepped down as departmental head to return to full-time teaching in 1979, the chemical engineering curriculum looked remarkably similar to that at the inception of his chairmanship. The lack of change did not represent passivity on the part of the department, however. Instead, following the old "if it isn't broke, don't fix it" dictum, the chemical engineering department of the 1970s refined and strengthened its curriculum by making changes to existing courses when necessary and by adding new elective courses to meet student needs and interests.

Chemical Engineering Curriculum, 1979–1980

Freshman Year

Fall	**Spring**
English 1013, Composition	English 1023, Composition
Chem. 1125, College Chemistry	Math. 2565, Calc. and Analytic Geometry
Math. 2555, Calculus and Analytic Geometry	Humanities-Social Sciences Elective

G.E. 1122, Engineering Graphics I	Ch.E. 2123, Chemical Engineering I
G.E. 1000, Engineering Orientations	Elective

Sophomore Year

Fall	**Spring**
Phys. 2053, University Physics I	Phys. 2063, University Physics II
Phys. 2071, Univ. Physics Lab I	Phys. 2081, Univ. Physics Lab II
Math. 2573, Calculus and Differential Equations	Humanities-Social Sciences Elective
Ch.E. 2133, Chemical Engineering II	Chem. 3614, Organic
Chem. 3604, Organic	Math. 3403, Differential Equations
Elective	E.S. 2003, Statics

Junior Year

Fall	**Spring**
Ch.E. 3233, Unit Operations I	Ch.E. 3243, Unit Operations II
Ch.E. 3312, Chemical Engineering Laboratory I	Ch.E. 3322, Chem. Engr. Laboratory II
Chem. 3504, Physical	Chem. 3514, Physical
Ch.E. 4533, Thermodynamics	Econ. 2013, Principles
E.S. 3003, Dynamics	Ch.E. 4543, Phys. and Chemical Equilibria
Humanities-Social Sciences Elective	Ch.E. 4411, Seminar

Senior Year

Fall	**Spring**
Ch.E. 4433, Design I	Ch.E. 4433, Design II
Ch.E. 4133, Chemical Processes and Kinetics	Technical Elective (6 hrs.)
Ch.E. 3332, Chemical Engineering Laboratory III	Humanities-Social Sciences Elective (6 hrs.)
Ch.E. 3253, Unit Operations III	E.E. 3903, Electric Circuits
E.S. 3103, Mechanics of Materials	
Ch.E. 4632, Projects	

During the seventies, the U of A Department of Chemical Engineering never lost sight of its original goal of quality undergraduate instruction, but it did reflect an increased focus on research. Although funding for academic research throughout the United States had increased significantly since the end of World War II, the year 1970 marked a turning point for sponsored projects. Support from all sources began to increase phenomenally during that year, and, as the money-flow sped up, so too did research.

In general, chemical engineering research during this era began to be aimed at a new target audience for approval. In the past, chemical engineers had gauged their efforts by the appeal their research had for other chemical engineers, industry, and federal and state governments. By 1970, the general public, reflecting an interest in the environmental and social impact of the chemical process industries, began to take a keen interest in chemical engineering developments. Besides the interest the public took in their research, chemical engineers themselves shifted their focus from empirical to more theoretical approaches, all the while keeping in mind the practical applications of their projects. Research interests for chemical engineers also spread out in a number of different directions.

This diversification of research interests can be seen in the projects that chemical engineering faculty members worked on during the 1970s. Louis Thibodeaux studied the composition of industrial wastewater, modeled pollution from point sources, and worked on a packed crossflow cascade tower. Charles Springer looked into interfacial tension in fluids. Robert MacCallum studied drying methods of biomaterials that could be used to replace herbicides. Jim Turpin researched washing techniques that would use less water and non-Newtonian fluid flow in pipes. James Couper studied methods of improving yields of ethyl acrylate. Robert Babcock investigated methods of lowering potassium and sodium contents of whey. Charles Thatcher delved into computer simulation and control of chemical processes to be used in teaching.

In a 1979 essay in *Arkansas Engineer,* Louis Thibodeaux addressed the "Environmental Revolution" and its impact on chemical engineering research. Traditionally, Thibodeaux noted, chemical engineers "have been educated to produce large quantities of substances originating from the industrial and university laboratories of the chemist." However, with the passing of years and the realization that there were unforeseen consequences of their past work, chemical engineers of the seventies and beyond had to acknowledge that "the same technical and scientific knowledge that fostered the creation and production of chemical substances must be directed to minimizing the risks thereof."

To address the issues raised in Thibodeaux's article, several members of the chemical engineering faculty combined their efforts to create the Chemical Hazards Research Laboratory (CHRL). Although it had existed informally for a few years before, the CHRL became an officially recognized part of the University of Arkansas in February 1979. It was staffed by four faculty members, seven graduate students, and three fifth-year seniors and was funded by five federal grants with supplementary funding from the state.

Chemical engineering faculty and students used the CHRL to conduct research into a number of areas. Charles Springer and senior chemical engineering student Joe Ryburn worked on "computer simulation models to predict the concentration of hydrazine and related compounds downwind from spills." Jerry Havens worked with undergraduate and graduate students on models of "fire and explosion hazards associated with fuels and chemicals." Charles Thatcher researched "the magnitude of the trickle flow of water in sediment." Graduate student Patty Christy and Professor Louis Thibodeaux investigated "synthesizing a chemodynamic model that will predict the waterborne hazards associated with the spill of sinker chemicals." Thibodeaux also teamed with graduate students Richard Dickerson and Harold Heck and Civil Engineering Professor David Parker in researching "conventional industrial wastewater treatment processes as sources of air pollutants," and he worked with graduate student Karen Kuhn and a number of undergraduates in studying "chemical transport processes within sediment beds of streams."

Increases in research did not necessarily correlate into increases in space. Unlike the 1960s, which saw the opening of the Science-Engineering Center that alleviated some of the overcrowding in Engineering Hall, and the 1980s, which witnessed the creation of the massive Bell Engineering Center, the Department of Chemical Engineering did not directly benefit from any of the university's building activities during the 1970s. As usual, the need for more space was acute; by the end of the decade, plans were being created to provide it. In the meantime, though, the department had to settle for what it had: overcrowded conditions by the mid-1970s with no money for new buildings because most of the Engineering College's extra funds paid for faculty raises and maintenance and operations costs during a time of rampant inflation.

Chemical engineering students—indeed, all university students—

benefited from new construction on the U of A campus, though. For students, the center of campus activity became the new Arkansas Union, completed in 1973 at a cost of $6.5 million. That same year, the $3.5 million Communications Center (today's Kimpel Hall) opened. By the end of the decade, the university also had built new Business Administration and Plant Sciences buildings and had made additions to the Fine Arts Center, Waterman Hall, Barnhill Fieldhouse, and Razorback Stadium. Memorial Hall had been renovated, and plans for a facelift for Old Main were underway.

Although it continued to be the Engineering College's top outreach program to the rest of the university and to the Fayetteville community, Engine Week, as Engineers' Week was known by the 1970s, faced the same sorts of difficulties as it had during the previous decades. Due partly to the increase in the number of extracurricular activities available to engineering students and also to the difficulties inherent in the engineering curricula, a general decline in interest in Engine Week became evident by the mid-seventies. "Engine Week functions no longer draw even adequate support from students," noted an *Arkansas Engineer* editorial. The waning interest among students only compounded the problem of rising costs for the week's activities. The same *Arkansas Engineer* editorial warned that the time had come to either put up or shut up: engineering students needed to make the week more appealing and cost-effective or face the possibility of ending some or all of the activities.

Besides the lower interest students had in Engine Week, the Engineers' Rally continued to draw the ire of some spectators. One observer noted in 1971 that the rally had become "long on grossness and short on humor and talent." As student and faculty dissatisfaction with the event grew, the Engineers' Rally increasingly faced the fate of other campus activities that had recently fallen by the wayside.

Chemical engineering students continued to participate in the activities of professional and honor societies during the seventies. The newest of the honor societies, Omega Chi Epsilon, came to the University of Arkansas in 1970. This national chemical engineering honor society chartered the Sigma Chapter on December 14 of that year. The Sigma chapter had previously been known as Tau Sigma, a local chemical engineering society. In 1975, the Society of Women Engineers, reflecting the jump in women enrolling in the College of Engineering, became the first professional society at the University of Arkansas designed specifically for women engineers.

Besides the new societies, older ones such as the AIChE and Alpha Chi Sigma continued to garner attention and to evolve in new directions. Changes in membership also became evident during the 1970s; women and minority students increasingly played important roles in the societies. T. A. Walton, a black student, became the first minority chemical engineer to serve as vice president of the AIChE and as Tau Beta Pi pledge trainer late in the decade.

Engineers' Day, 1974. Candidates for St. Patrick, *l. to r.:* David Rowe, Agricultural Engineering; Bob Holt, Chemical Engineering; Andy Wood, Mechanical Engineering; Brian Foster, Industrial Engineering. (*Arkansas Engineer,* April 1974)

The 1970s marked the beginnings of significant technological developments that changed the way university students participated in the learning process. Although computers had been on campus since the early 1960s, problems of access and user-friendliness kept many students from using them during the seventies. The new electronic calculators that became available during the decade, however, had a tremendous impact on the lives of students taking mathematics and scientific—and, of course, engineering—classes, as is evident in the following paragraph from the *Arkansas Engineer:*

> During the recent school year engineering students have been making changes in the manner in which they arrive at numerical solutions to their homework and test problems. The twelve inch slide rule that once swung from the belt of the book-laden engineering student has begun to give way to the palm sized electronic calculator. Over seventy-five different models of portable electronic calculators are available, and their list prices range from $69.95 to $995.00. The lower priced models are capable of performing only the basic arithmetical functions (add, subtract, multiply, and divide), while the higher priced models perform some or all of the arithmetical calculations encountered in any engineering discipline.

Chemical engineering students also continued to participate in campus life. One of the highlights of the decade came in 1971 with the U of A's centennial festivities. Numerous events celebrated the milestone. In June, a Centennial Convocation was held at the National Cathedral in Washington, D.C., and a commemoration ceremony was held to honor the decision to locate the university in Fayetteville. In January 1972, faculty and students participated in a convocation to commemorate the hundredth anniversary of the first day of classes.

For the sports enthusiast, 1969 through 1979 continued the rich traditions begun in earlier decades. During the first year of Couper's tenure as department head, the Razorback football team participated in what has been variously termed the "game of the century" or the "Great Shootout"; the 1969 game versus Texas was attended by President Richard Nixon and ended in a 15–14 defeat for the Razorbacks. Under head coach Lou Holtz during the mid- and late seventies the team achieved a level of national prestige that rivaled that of the sixties. Between 1974 and 1979, the team finished in the national top ten four times.

Dickson Street, early 1970s. (Courtesy Joe Neal / Shiloh Museum of Ozark History)

The basketball team, under Coach Eddie Sutton, began to rival the football team in terms of recognition by the late 1970s. The Razorback roundballers finished in the top ten three years in a row, won three

Southwest Conference championships, and made it to their first NCAA Tournament Final Four. Between 1976 and 1979, the basketball team won more games (eighty-three in all) than any other team in the nation. For the 1977–1978 school year both the football and basketball teams ranked third at the end of their respective seasons, with baseball ending up sixth.

Other Razorback teams also attained unparalleled heights during the seventies. The tennis team finished among the top ten on three separate occasions. In 1979, the baseball team reached the finals of the College World Series. The cross-country team won its fourth consecutive Southwest Conference championship in 1977.

Students of the late sixties and seventies enjoyed hanging out on Dickson Street, and, if they had cars, could often be seen cruising from the Vic-Mon to the A&W on College Avenue. Further north on College, they could watch movies at either the 71 Drive In or the 62 Drive In. On Dickson Street, the hot spots included the ROTC and the D-Lux, George's and the Library, the Whitewater Tavern and the Minute Man burger house. Later in the seventies, the Y'all Come Back Saloon on College entertained Fayetteville's urban cowboys. Fayetteville in the seventies also boasted two miniature golf courses, one next to the Huddle Club on College, and the Hillbilly Holler, just out of town to the north. The Northwest Arkansas Mall opened in 1972.

In 1977, the Department of Chemical Engineering celebrated its seventy-fifth anniversary as a part of the University of Arkansas. The celebration included a series of lectures presented by distinguished alumni, including Ray Adam and Robert N. Maddox. James R. Fair, director of engineering technology for Monsanto, and Margaret Skillern, director of the American Institute of Chemical Engineers, also delivered lectures.

When James Couper stepped down as department head three years later, he could look back on his chairmanship as a period of remarkable stability. The tenure of James Gaddy, Couper's successor, would not be as stable, but it would be just as successful.

Biographical Sketches, 1969–1979

Charles Thatcher became Alcoa Distinguished Professor of Chemical Engineering in 1970. Before his arrival at the University of Arkansas, Thatcher had served as head of the Department of Chemical Engineering at Pratt Institute, Brooklyn, New York, where he also served eight years as dean of engineering. Prior to his years at Pratt, Thatcher taught at the University of Michigan, where he had received his B.S.E., M.S.E., and Ph.D. degrees.

Charles Thatcher. (ChemE collection)

Thatcher came to the U of A already well known as the author of the *Fundamentals in Chemical Engineering* textbook, and over the next twenty-two years he became one of the most beloved and admired teachers within the department. He twice served as department head and was a founding member of the University of Arkansas Teaching Academy. He also was one of two faculty advisors for the U of A Sailing Club and served as international president of Sigma Chi.

"He was very . . . you might say disciplinarian," remembered a former colleague. "He was almost military-like in his focus. His mind was very active, he was very much—almost a renaissance man in the things he could converse in. A very interesting person."

Thatcher retired in 1992. In his honor, Dr. Tom Spicer, who knew Thatcher as both a teacher and a colleague, wrote the following poem:

Let me at 'em

To the man with the face that
sank the hopes of a thousand students

There's a certain enjoyment of life
to be seen in the sparkle of the eye
of a person committed to others
to get them to grow
to get them to search
to get them to try.
Not to be easy on anyone involved
especially those
pulled and stretched and tried.
But a joy—a vigor—a sparkle
to be kindled in another person's eye.

Jerry Havens arrived at the U of A as assistant professor of chemical engineering in 1970. Havens received his B.S.Ch.E. from the University of Arkansas, his M.S.Ch.E. from the University of Colorado, and his Ph.D. from the University of Oklahoma. He attained the rank of associate professor in 1975, professor in 1980, and Distinguished Professor of Chemical Engineering in 1987. He is director of the University of Arkansas Chemical Hazards Research Center.

Jerry Havens. (ChemE collection)

Distinguished Professor Havens is a registered professional engineer, has industrial experience with Phillips Petroleum and Procter and Gamble, and has served as an officer in the U.S. Army Chemical Corps.

Havens's primary research interests are in atmospheric dispersion of heavy gases and fire/explosion phenomena. He is internationally recognized as an expert in methodologies for predicting atmospheric dispersion of hazardous, denser-than-air gases, and has served as technical advisor and consultant to governmental and industrial agencies too numerous to list here. Over the course of his career, Havens has also compiled a lengthy list of publications, presentations, and research reports focusing especially on heavy gas dispersion and fire and explosion phenomena. He holds memberships in the AIChE, Sigma Xi, and the American Chemical Society.

Professor Havens conducts a dense gas dispersion research program at the U of A that has received approximately $8 million in funding. He has served on the Research Proposal Review Board of the Commission of European Communities Research and Development Directorate and was the only nonmedical doctor on a sixteen-member team from fourteen countries at the International Medical Commission on Bhopal in 1994. Havens also serves on the NOAA/National Ocean Service program review panel, the International Editorial Board of the Institution of Chemical Engineering journal, and the International Editorial Board of the *Journal of Hazardous Materials.*

Doctor Havens is regarded as one of the university's outstanding researchers, and he received the inaugural Award for Excellence in Research from the Arkansas State Board of Higher Education in 1988.

Robert N. MacCallum became an assistant professor of chemical engineering at the U of A in 1970, when he replaced Jim Turpin, who was on leave for the academic year. In 1975, he received a promotion to associate professor.

MacCallum earned his B.S.Ch.E. and M.S.Ch.E. from the University of Texas and his Ph.D. from Rice University. Prior to his arrival at Arkansas, he worked for the Humble Production and Research Company in Houston, Texas.

In 1975, MacCallum served as a part-time visiting professor at the University of Arkansas-Pine Bluff, where he not only taught but also provided career guidance for minority students who wished to enter engineering fields.

MacCallum left the U of A in December of 1979, but he has maintained contact with the department in the years since, especially while working at the Amoco Tulsa Research Center. In this position he served as team leader, and eventual supervisor, for a computer-aided process design group, and his role with Amoco led to the corporation's donation of the ASPEN process simulator to the Department of Chemical Engineering.

Robert MacCallum

MacCallum has also worked as a supervisor of computer research with the Tulsa Research Center, as a process engineer with Amoco Worldwide Engineering and Construction and Amoco Natural Gas Liquids Business Unit, and as a process engineer for Altura Energy, Ltd., known today as Occidental Permian, Ltd. In each of these positions, he has been able to use the experience he gained on the faculty at the U of A: "Because of my background in academia and in industrial research, I tend to encourage our looking at new processes. . . . I find it really interesting and enjoyable that my teaching experience helps me in my daily job—'Hey, this basic knowledge we teach in the classroom really works and is useful!'"

Chapter Seven

The Gaddy Years, 1980–1990

The 1980s brought a decline in funding for higher education in the United States; the drop was particularly serious in Arkansas. Despite its status as flagship for the University of Arkansas System, the Fayetteville campus did not escape the declines. "The most serious crisis in the recent history of the institution," wrote Chancellor Daniel Ferritor in 1987, "has resulted from repeated cuts to established budgets and a prolonged period when vital expenditures had to be deferred."

Although total enrollments at the U of A leveled off during the eighties, student numbers in career-oriented programs such as chemical engineering grew. In the College of Engineering, freshman enrollment increases of as much as 20 percent (in 1985) compounded the problem of lower state appropriations and led to an ever greater effort at attaining private funding of programs. The Department of Chemical Engineering, under the chairmanship of Professor James Gaddy after 1980, ranked in the top twenty nationally in terms of overall enrollment and in the top five for freshman enrollment by the mid-eighties. By the end of Gaddy's tenure, the department had the fifth largest undergraduate, the fourth largest master's, and the eighteenth largest doctoral enrollments among U.S. chemical engineering programs.

The growth was not simply a matter of quantity. Each enrolling class raised expectation levels for future successes. For example, the freshman class for 1987–1988 included twenty-two high school valedictorians, eight Governor's Scholars, and eleven National Merit semifinalists. The same year, the department's students received thirty College of Engineering Academic Scholarships, forty-two Combs or Freshman Scholarships, twelve national awards and scholarships, and four University awards. Fifty made the Dean's List, thirty juniors and seniors were named Academic All-Americans, and seventeen had a 4.0 GPA. The chemical engineering team also won that year's Engine Week Quiz Bowl.

The make-up of the chemical engineering student body continued to change during the Gaddy years. Although women, minority, and international students had made up small portions of the program's students for decades, the 1980s saw increases in all three of these groups. These changes were not confined only to the Department of Chemical Engineering, though; the entire college and, indeed, most U.S. engineering programs, saw the same sorts of changes during the eighties and into the nineties. By 1990, women received 15.4 percent of all engineering degrees in the United States, although they tended to specialize in different fields than men: 25 to 40 percent of the degrees granted in materials, systems, industrial, biomedical, environmental, and chemical engineering went to women, while the percentages were much smaller in electrical, mechanical, aerospace, petroleum, and marine engineering. The proportions were slightly different for minority engineers: Hispanic Americans were roughly proportional to other engineering students in terms of their specializations; African Americans showed a slight preference toward electrical and industrial engineering; and Asian Americans exhibited clear preferences for electrical engineering. At the U of A, the College of Engineering attracted significantly more black students than the rest of the university, thanks in large part to a Rockefeller Foundation grant designed for that purpose.

Building upon his predecessors' achievements, Gaddy steered the department to new heights of national and international visibility, stared down the decade's budget crises, and oversaw an increasingly complex network of chemical engineering teaching and research. Besides a new department head, the chemical engineering faculty changed significantly due to additions, retirements, and departures. Despite the changes, the stability that began during Philip Bocquet's tenure as department head, and that had continued during the Couper years, remained evident.

The faculty's dedication to its teaching mission increasingly led to honors and awards during the Gaddy years. Undergraduate students consistently ranked the department at the top of the College of Engineering on student evaluations and pregraduation surveys, with individual professors always being ranked from good to excellent. The high esteem students had for the faculty as a whole came despite the relatively poor student to teacher ratio of 25 to 1 within the department and faculty salary levels 20 percent lower than the national average. In 1983, the Ray C. Adam Chair for Young Faculty was established to honor young chemical engineering faculty who excelled in research and teaching.

The following faculty members arrived during the 1980s. Some came to fill positions opened by the retirement or departure of earlier members; others came to fill newly created positions.

1. James L. Gaddy. Professor of Chemical Engineering (1980–); Head of the Department of Chemical Engineering (1980–1989); Distinguished Professor of Chemical Engineering (1987–1992;); Distinguished Professor Emeritus (1992–). B.S.Ch.E., Louisiana Polytechnic University; M.S.Ch.E., University of Arkansas; Ph.D., University of Tennessee.

2. James E. Halligan. Professor of Chemical Engineering (1979–1985); Dean of the College of Engineering (1979–1983); Director of the Engineering Experiment Station (1979–1983); Vice Chancellor for Academic Affairs (1983–1984); Interim Chancellor (1984–1985). B.S.Ch.E., M.S.Ch.E., Ph.D., Iowa State University.

3. Edgar C. Clausen. Associate Professor of Chemical Engineering (1982–1986); Professor (1986–). B.S.Ch.E., M.S.Ch.E., Ph.D., University of Missouri-Rolla, P.E.

4. J. Reed Welker. Professor of Chemical Engineering (1983–). B.S.Ch.E., M.S.Ch.E., University of Idaho; Ph.D., University of Oklahoma.

5. David W. Suobank. Assistant Professor of Chemical Engineering (1985–1986). B.S.Ch.E., University of Arkansas; Ph.D., California Institute of Technology.

6. Thomas O. Spicer, III. Visiting Assistant Professor of Chemical Engineering (1984–1986); Assistant Professor (1986–1988); Associate Professor (1988–1996); Professor (1996–). B.S.Ch.E., M.S.Ch.E., Ph.D., University of Arkansas; P.E.

7. Alfred A. Silano. Visiting Research Professor of Chemical Engineering (1985–1987); Research Professor of Chemical Engineering (1988–present). B.S., Rutgers University; M.S., New Jersey Institute of Technology; Ph.D., Rutgers University.

8. William A. Myers. Instructor in Chemical Engineering (1956, 1985–). B.S.Ch.E., M.S.Ch.E., University of Arkansas; P.E.

9. John P. Middleton. Adjunct Professor of Chemical Engineering (1987–1990). B.S.Ch.E., University of Arkansas.

10. Michael Dean Ackerson. Assistant Professor of Chemical Engineering (1988–1993); Associate Professor (1992–). B.S.Ch.E., M.S.Ch.E., University of Missouri-Rolla; Ph.D., University of Arkansas.

11. Richard K. Ulrich. Assistant Professor of Chemical Engineering (1987–1991); Associate Professor (1991–1995); Professor (1995–); Professor of Electrical Engineering and Chemical Engineering (1999–). B.S.Ch.E., University of Texas; M.S.Ch.E., University of Illinois; Ph.D., University of Texas, P.E.

12. W. Roy Penney. Professor of Chemical Engineering (1989–); Head of the Department of Chemical Engineering (1990). B.S.M.E., M.S.M.E., University of Arkansas; Ph.D., Oklahoma State University; P.E.

These new faculty additions played crucial roles in the department's curricular developments. Across the nation, chemical engineering departments in the 1980s (and into the 1990s) developed curricula that placed more stress on communications and management skills while continuing to emphasize environmental and social issues.

Chemical Engineering Curriculum, 1980–1981

Freshman Year

Fall	Spring
MATH 2555, Calculus I	MATH 2565, Calculus II
CHEM 1125, College Chemistry	CHEG 1124, Chemical Engineering I
ENGL 1013, Composition	ENGL 1023, Composition
GNEG 1122, Engineering Graphics	Humanistic-Social Sciences Elective
GNEG 1001, Engineering Orientation	

Sophomore Year

Fall	Spring
MATH 2573, Calculus III	MATH 3403, Differential Equations
CHEM 3604, Organic	CHEM 3614, Organic

PHYS 2053, University Physics I	PHYS 2063, University Physics II
PHYS 2071, Univ. Physics Lab I	PHYS 2081, Univ. Physics Lab II
CHEG 2233, Momentum Transport	CHEG 2533, Thermo. of Single-Comp. Sys.
CHEG 2311, Chemical Engineering Lab I	Humanistic-Social Sciences Elective

Junior Year

Fall	**Spring**
CHEM 3504, Physical	CHEM 3514, Physical
EGSC 2003, Statics	EGSC 3103, Mechanics of Materials
CHEG 3243, Heat Transport	CHEG 3133, Chem. Processes and Kinetics
CHEG 3322, Chemical Engineering Lab II	CHEG 3253, Mass Transport I
CHEG 3543, Thermo. of Multi-Component Systems	CHEG 3621, Chem. Engr. Seminar
Humanistic-Social Sciences Elective	Humanistic-Social Sciences Elective

Senior Year

Fall	**Spring**
CHEG 4273, Mass Transport II	CHEG 4332, Chemical Engineering Lab III
CHEG 4433, Design I	CHEG 4433, Design II
CHEG 4632, Projects	ELEG 3903, Electric Circuits
Technical Electives	Technical Electives
Humanistic-Social Sciences Elective	Humanistic-Social Sciences Elective

Notably different from the year before, the 1980–1981 curriculum dropped the three courses titled Unit Operations and replaced them with Heat Transport, Mass Transport I, and Mass Transport II. The senior-year technical electives consisted of twelve hours of coursework, approved by the department. Beginning in the 1986–1987 school year, the department created seven technical options in chemical engineering. These options, to be fulfilled through the technical elective courses, included biotechnology, semiconductor, control and safety, simulation and optimization, materials science, specialty chemicals, and environmental subspecialties. Other than these changes, the curriculum remained stable during the Gaddy years. Faculty members constantly refined individual courses, however, to meet contemporary demands of society and industry.

As the availability of computers for classroom, personal, business, and industrial use increased during the eighties, the integration of computers into instruction became of primary importance for all engineering educators. "We have constantly updated our curriculum," noted U of A engineering dean Neil M. Schmitt in 1983, "to ensure computer literacy and to use the computer as a teaching and design tool."

By the mid-1980s, all chemical engineering courses included at least some computer work. A master computer programs, PROPS, provided the department with physical property data and simulation of chemical engineering equipment for instructional and experimental purposes. During the 1987–1988 school year, James Couper, after studying numerous other undergraduate and graduate programs' design curricula, revised the chemical engineering design courses to include the most up-to-date computer-aided design technology. The same year, Richard Ulrich added instruction in FORTRAN and utility programs to his mass transfer courses; Ulrich and Couper introduced process models such as ASPEN, RUNDIST, and FLASHER into their courses; Michael Ackerson wrote software for modeling basic processes in his thermodynamics courses; and Ackerson and Tom Spicer introduced TK Solver Plus for solving complex calculations in their classes.

In addition to the ever-increasing emphasis on computer literacy, the College of Engineering decreed in 1985 that a new pre-professional program of study had to be completed before a student would be allowed to enroll in any 2000-level or higher engineering course. The new program required students to complete three hours of composition, three hours of technical composition, five hours of Calculus I and II, and five hours of college chemistry and its laboratory *or* eight hours of general chemistry and its laboratory.

During the mid-eighties, the department began providing tutors for students in each of the chemical engineering courses offered. The tutors, coupled with the availability of teaching assistants on the new freshmen chemical engineering floors in Yocum and Humphreys dormitories, provided valuable help to students engaged in the sometimes overwhelming chemical engineering curriculum.

The end of the Gaddy years brought a few minor changes to the U of A curriculum, most notably in the form of four new upper- and graduate-level courses— Chemical Engineering Reactor Design (a required course of all juniors), Microelectronics Fabrication and Materials,

Biochemical Engineering Fundamentals, and Chemical Process Safety. These courses were actually added during the transitional period between the chairmanships of Gaddy and Robert Babcock. The final chemical engineering curriculum before the Babcock years looked like this:

Chemical Engineering Curriculum, 1989–1990

Freshman Year

Fall	Spring
MATH 2555, Calculus I	MATH 2565, Calculus II
CHEM 1123, University Chemistry	CHEG 1123, Intro. to Chem. Engineering II
CHEM 1131, University Chemistry Lab	ENGL 1023, Technical Composition
ENGL 1013, Composition	Humanistic-Social Elective
CHEG 1113, Intro. to Chemical Engineering I	CHEG 1212, Chemical Engineering Lab I

Sophomore Year

Fall	Spring
MATH 2573, Calculus III	MATH 3403, Differential Equations
CHEM 3604, Organic Chemistry	CHEM 3614, Organic Chemistry
PHYS 2053, University Physics I	PHYS 2073, University Physics II
PHYS 2061, University Physics Lab I	PHYS 2081, University Physics Lab II
CHEG 2133, Momentum Transport	CHEG 2313, Thermo. of Single-Comp. Sys.
Humanistic-Social Sciences Elective	Humanistic-Social Sciences Elective

Junior Year

Fall	Spring
CHEM 3504, Physical Chemistry I	CHEM 3514, Physical Chemistry II
MEEG 2003, Statics	MEEG 3013, Mechanics of Materials
CHEG 3143, Heat Transport	CHEG 3333, Chem. Engr. Reactor Design
CHEG 3322, Chemical Engineering Lab II	CHEG 3153, Mass Transport I
CHEG 3323, Thermo. of Multi-Component Systems	ECON 2013, Principles of Economics
Humanistic-Social Sciences Elective	

Senior Year

Fall	Spring
CHEG 4163, Mass Transport II	CHEG 4332, Chemical Engineering Lab III
CHEG 4413, Design I	CHEG 4443, Design II
CHEG 4423, Automatic Process Control	ELEG 3903, Electric Circuits
Technical Electives	Technical Electives
CHEG 3221, Chemical Engineering Seminar	Humanistic-Social Sciences Elective

Despite a significant decrease in job openings for engineers in general, and chemical engineers in particular, during the 1980s, graduates of the U of A Department of Chemical Engineering consistently received offers. Indeed, by the late eighties, the department's graduates were receiving twice as many employment offers, and higher starting salaries, than the national average. Chemical engineers also received the highest starting salaries ($33,500 for undergraduates in 1989) of all U of A graduates.

The job offers reflected the department's shifting emphasis in terms of its teaching and research expertise—while employment opportunities leveled off in some of the more traditional, process-oriented fields of chemical engineering, job openings in newer fields like biotechnology, semiconductor development, and hazardous waste treatment more than made up for the losses. And it was through research and expertise in those newer fields that the department developed a growing reputation for quality. Thus, employers viewed U of A chemical engineering students as exceptionally well qualified in the fields with the highest demand.

As had been the case for the preceding decades, research played an ever-larger role in departmental life under Gaddy's leadership. The research endeavors of chemical engineering faculty, graduate students—even undergraduates—emphasized the practical, hands-on knowledge that would be necessary for student success in the "real world." Research was a primary means of education, but it also brought publicity and renown to the department, the College of Engineering, and the U of A as a whole.

Gaddy's arrival sparked an even greater emphasis on research quantity and quality. Among the research projects during Gaddy's early years, a three-and-a-half-year project for the U.S. Coast Guard stood out. Under the direction of Jerry Havens, the project investigated the dispersal of heavy gases such as propane, methane, ammonia, and chlorine, following accidental releases. The project was funded at $400,000 and provided research opportunities for five graduate students.

By 1986, the U of A Department of Chemical Engineering ranked sixth in the nation in externally funded research, focused especially on three "centers of excellence": biochemical engineering, heavier-than-air gas disposition, and chemical hazards management. With the arrival of new faculty members came new and ever more diverse research endeavors, and new graduate students from across the United States, even from around the world, were attracted to the program and its centers of excellence. The department's faculty generally published and presented more than 150 papers per year.

The *Annual Report* of the University for the 1986–1987 school year noted the following areas of research excellence within the department:

1. *chemical hazards:* Under Professors Havens, Welker, and Spicer, chemical hazards researchers focused on "the prediction of hazards that could result in catastrophic accidents and the prevention of such accidents." The Department of Chemical Engineering had the only laboratory in the nation for experimental measurement of dense gas dispersion.

2. *biotechnology:* Under Professors Gaddy, Clausen, and Eldridge, biotechnology research focused on developing "methods to convert coal directly into fuels and chemicals."

3. *tertiary oil and tar sand:* Under Professors Babcock and Turpin, tertiary oil and tar sand research investigated "methods to recover oil from depleted reservoirs or heavy oil and tar deposits."

4. *environmental science:* Professors Springer and Turpin led investigations into "pollution aspects of various toxic emissions."

5. *process simulation and optimization:* Professors Thatcher and Gaddy researched the creation of computer simulation models and the study of alternative operation strategies to improve "economy in chemical processes through reduced raw materials use and lower energy consumption."

6. *process safety:* Under Professors Springer and Welker, investigations into process safety sought to develop "procedures to evaluate and minimize the risks associated with the manufacture, storage, and transportation of large quantities of flammable or toxic chemicals."

7. *materials science:* Professor Couper led research "to obtain a better understanding of the performance and properties of plastics."

The department's growing renown as a center of chemical engineering research began to attract international attention. Notably, researchers from around the world traveled to Fayetteville to take advantage of the faculty's expertise and the department's laboratory resources. The outside researchers came from all corners of the globe and included distinguished investigators in their own right, such as David MacKay of the Technical University of Nova Scotia, Xin Fei of the Beijing Institute of Control Engineering, Junjie Wu of the East China Institute of Chemical Technology, Liu Rui-Qin of the Institute for Microbiology Academia Sinica, Cervasio Antorrena of the University of Santiago, and Ali el-Shafei of the Egyptian National Research Center.

Another reflection of the department's growing international reputation came in the form of exchange programs with other universities. In 1987, the Department of Chemical Engineering was chosen by the Swedish government for the CHUST Fellowship program. The program sponsored "Sweden's most outstanding students for rigorous courses of study and research at strong universities in the United States."

By the end of the Gaddy years, the Department of Chemical Engineering had moved into fifth place among U.S. departments in terms of externally funded research. Its national and international renown would continue to blossom during the 1990s. But not all of the department's acclaim resulted from the individual initiative of faculty members; new facilities also increased the U of A chemical engineering program's prestige.

Not since the Barker years had the Department of Chemical Engineering been the direct beneficiary of the university's building program. For three decades, chemical engineers had been relegated to a wing of Engineering Hall that, although modern and sufficient when it opened in the 1950s, severely lacked space and up-to-date laboratories by the beginning of the Gaddy era. The 1980s, however, brought about a drastic change in the department's physical environment.

The first major change came in the early eighties, when the university instituted a comprehensive plan to create a new off-campus engineering research and education station. A Legislative Study council recommended the new complex, soon dubbed "Engineering South," to "stimulate technological development within the state by providing research space and facilities for a more viable Engineering Experiment Station." Located approximately two miles south of the main Fayetteville campus, Engineering South developed in two stages: the first priority was to remodel and renovate the nearly 200,000 square feet of existing buildings on the property (it had formerly been the site of Bear Brand Hosiery, Inc.); the second phase would come with the construction of new buildings for additional space. Professor Robert Babcock coordinated much of the planning in his capacity as liaison between the architects and the university administration. When all the planning was finished, the Department of Chemical Engineering received 31,040 square feet of new space for research and teaching.

Engineering South. (ChemE collection)

The second major development for the physical environment during the eighties came about not only because of a concern with overcrowded classrooms and lack of laboratory and office space, but also because of the need for the College of Engineering to upgrade its facilities for reaccreditation by the Accreditation Board for Engineering and Technology. To meet this necessity, the college put forth a three-prong attack: obtain and renovate Engineering South, renovate Engineering Hall, and, most important, build a new engineering classroom, laboratory, and office building on campus.

Like the Chemistry Building basement, the Barn, and the Engineering Hall Annex, the new building would become the domain of the chemical engineers. Dean of Engineering James Halligan, a chemical engineer by training, ordered excavations to begin in 1981, putting pressure on the state to provide full funding for the building. The actual construction started the following year. The building—named the Bell Engineering Center in honor of Owen and Hildur Bell, parents of Melvyn L. Bell (B.S.E.E., 1960), who donated $8 million for the building's equipment and furnishings—opened in January 1987. The 200,000 square-foot building's central space featured a sky-lighted atrium, running southeast to northwest. It's lower level contained utility, mechanical, and service/support facilities. A two-story auditorium provided ample room for classes as large as three hundred. Besides new classrooms and laboratories for chemical, industrial, electrical, and civil engineering, the Bell Center contained office space for faculty, graduate students, and the College of Engineering administration. A comprehensive computer network linked the new center to Engineering Hall and the Mechanical Engineering Building, and elevated walkways connected it physically to Engineering Hall and the Science-Engineering Building.

Bell Engineering Center gave the Department of Chemical Engineering new undergraduate laboratories and allowed for the expansion of both classroom and laboratory curricula. Additionally, U of A chemical engineering alumnus Robert N. Maddox, who retired in 1987 from Oklahoma State University's Department of Chemical Engineering after thirty-six years, established the Maddox Student Reference Fund and donated his personal library to create the Maddox Student Reference Center on the Bell Center's second floor. The new reference center became a valuable tool for both chemical engineering students and faculty and contains a wealth of information for everyone associated with the department.

Bell's second floor also housed the chemical engineering lounge, which quickly became a prime spot for socializing and studying. It was "the site of many an 'all-nighter,'" remembered one former student. "Toward the end of each semester, familiar faces would reappear as we put the final touches on the final project for the classes we were taking at the time."

When not taking classes or hanging out at Bell Engineering Center,

Bell Engineering Center—exterior. (ChemE collection)

chemical engineering students continued to take part in a variety of extracurricular activities. Engine Week reached a low point during the mid-1980s, when its festivities completely ceased due to lack of interest on the part of students and the general public. After a two-year hiatus, engineering students in 1988 created a central planning committee to

Bell Center—interior. (ChemE collection)

oversee a revival of the week. What resulted was a much more subdued Engine Week, without the pomp and circumstance of earlier events, but with enough student involvement to sustain it.

Taking the place of the skits and dances that had fallen by the wayside, activities such as T-shirt design contests, picnics, volleyball tournaments, tug-of-wars, and faculty dunking booths brought engineering students a more relaxed environment and lessened dependence on outside sources of support. In 1989, a chemical engineering team consisting of Colonel William A. Myers, Eryk Hargrove, Wally Williams, and Joey Jacobs won the Engine Week Quiz Bowl, another new event, beating out another chemical engineering team headed by Dr. J. Reed Welker.

The U of A chapter of AIChE continued to play a substantial role in the chemical engineering department's activities. Along with Omega Chi Epsilon, AIChE routinely garnered the support of at least three-quarters of the chemical engineering students, and it consistently won the Outstanding Chapter Award from the national organization. The chapter also sponsored extracurricular events for the benefit of both undergraduate and graduate students. For example, in the 1985–1986 school year, AIChE sponsored twenty-one events, including technical, scientific, and professional seminars, career guidance workshops, social events, and field trips. The organization also hosted the Mid-America Regional Student Chapter Conference in 1987, created and presented a thirty-minute slide show entitled *Chemical Engineering at the U of A* to high school students around the state, and conducted tours of the Fayetteville campus for prospective students and their parents. The AIChE also sponsored picnics that "provided a great opportunity for social interaction between the faculty, graduate students, and

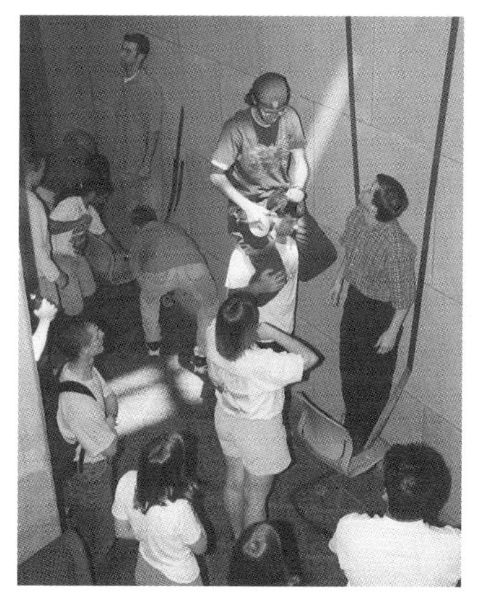

Engine Week, 1990s. Chemical engineering students are taping Dr. Bob Beitle to the wall. If he remains attached for thirty minutes, they will win.

undergrads ranging from Freshmen to 5th [year] seniors," remarked one student.

By the early 1980s, the Alpha Chi Sigma chapter at the U of A had fallen on hard times. On member recalled: "There was no money, the meetings were few, the regalia was tattered and incomplete; there was not really enough of us to ever have a proper initiation." To rectify the situation, a small group of students began to seek ways to renew interest in the organization. Ernie Lucas and John Meazle attended the national Alpha Chi Sigma Conclave at Cornell University for ideas on how to revive the chapter, and, soon thereafter, the group began to concentrate on holding regularly scheduled meetings and on actively recruiting qualified students. A strong core of pledges developed, and, thanks to their hard work over the next few years, the U of A chapter earned the distinction of top non-house chapter in the nation in 1985.

With Professor Gaddy's arrival came a new tradition for chemical engineering students and faculty to participate in, the Spring Awards Banquet. The first such banquet took place in 1980; they continue to be yearly highlights for the department today. The goal of the banquets was to take at least part of the place of the Engineering Week festivities that had fallen on hard times. To that end, the banquets featured not only award ceremonies but also skits and mock presentations from students to teachers,

Chemical Engineering Banquet. (ChemE collection)

while distinguished alumni provided the keynote address. To this day, the banquets provide a welcome respite for students and faculty under the stress of the year's end.

In May 1988, the Department of Chemical Engineering held an alumni reunion for all of its graduates through 1959. Nearly one hundred alumni and their spouses attended the event, which included presentations of faculty research, curriculum, and facilities. The event was held in conjunction with the department's Spring Awards Banquet.

Much had changed since these alumni left the U of A campus—new buildings, new classes, new faculty. Yet much remained the same: the Department of Chemical Engineering remained dedicated to providing a quality education to all of its students and to furthering the chemical engineering profession through its research activities. These commitments remained steady even after Jim Gaddy stepped down as department head in 1989. After a transition period, during which Charles Thatcher served as acting department head, Professor W. Roy Penney was hired in 1990 to chair the department; but, after discovering that his administrative duties required time that he would rather spend teaching and researching, Penney stepped down. His replacement, Robert Babcock, provided leadership for the final eleven years of the department's first century.

Biographical Sketches, 1980–1990

James L. Gaddy came to the University of Arkansas in 1980 to fill the shoes of Jim Couper as chair of the Department of Chemical Engineering. Gaddy would serve in that capacity, and as professor of chemical engineering, for nine years, stepping down as chair in 1989. He presided over a period of phenomenal change for the department, in terms of faculty development, physical environment, and national recognition.

Professor Gaddy received his B.S.Ch.E. from Louisiana Polytechnic University, his M.S.Ch.E. from the University of Arkansas, and his Ph.D. from the University of Tennessee. Prior to his return to the U of A in 1980,

Gaddy was a member of the chemical engineering faculty at the University of Missouri-Rolla and, before that, he was an instructor and NASA fellow with the University of Tennessee. As a faculty member at the U of A, Gaddy has been not only professor and head of the chemical engineering department, but also distinguished professor and distinguished emeritus professor.

Outside of the classroom, Gaddy has worked for Ethyl Corporation, Arkla Chemical Corporation, and, since 1992, he has been president of Bioengineering Resources, Inc. His professional and honorary affiliations include the AIChE, the American Chemical Society, the American Society for Engineering Education, Tau Beta Pi, Omega Chi Epsilon, Sigma Xi, and the American Association for the Advancement of Science. He is a registered professional engineer in Arkansas.

James Gaddy. (ChemE collection)

Gaddy has edited, authored, or jointly authored twenty-two books or monographs, 209 articles in journals or proceedings, and twenty-six research reports. He has taken part in more than 240 presentations to scholarly or professional meetings and has participated in more than one hundred lectures or seminars. He holds four U.S. patents and has two pending. Over the years he has served as a consultant to more than a dozen companies, and he has received such honors as being named Tau Beta Pi Eminent Engineer. His sponsored research projects have been funded to more than $20 million.

According to one source, Gaddy had a penchant for sardines, which he would devour in his office after lunch-time basketball games. His colleagues on the faculty also recall that, although he would not donate blood during the annual Red Cross Blood Drives, Gaddy always ate the cookies the others received.

J. Reed Welker came to the Department of Chemical Engineering as a professor in 1983. Welker received his B.S.Ch.E. and M.S.Ch.E. from the University of Idaho and his Ph.D. from the University of Oklahoma. His previous academic experience included stints as research engineer and associate director of the Flame Dynamics Laboratory at the University of Oklahoma Research Institute and as an instructor at the University of Idaho. For more than two decades prior to his arrival at the U of A, Welker had also worked in industry—first as a group and section leader with Oil Recovery Corporation; then as a senior staff member, project director, and vice president with University Engineers, Inc.; and finally as president of Applied Technology Corporation.

Professor Welker is remembered as a rigorous but fair teacher by his former students. "He was tough, but you learned a lot," recalled one. Welker has taught Thermodynamics I, Thermodynamics II, Introduction to Chemical Engineering I, Introduction to Chemical Engineering II, Mass Transfer I, Mass Transfer II, Chemical Process Safety, Chemical Engineering Lab III, Reactor Design, Transport I, Equilibrium Stage Operations, Research Proposal Seminar, and Graduate Seminars.

J. Reed Welker. (ChemE collection)

Welker's research interests include liquefied gas technology, fire behavior, risk analysis, heat transfer from fires, fire-control techniques, ignition of combustible materials, and chemical process safety. He has published thirteen books or chapters and fifty-one articles in professional journals or proceedings, and he has presented his findings at fifteen professional meetings and twenty-five invited seminars. Welker has also joined with Professor Charles Springer to present continuing education courses on chemical process safety.

Welker's national service includes work on the editorial

boards of professional journals, plus committee and chairmanship work for groups such as the American Gas Association, the AIChE, the National Academy of Sciences, and the National Research Council. At the University of Arkansas, Welker has kept himself busy as a member of the departmental graduate program committee, the departmental library committee, the committee on committees, the honorary degree committee, the scholarship committee, and the English as a Second Language committee. He has also served as chemical engineering safety officer.

The value of Welker's research is attested to by the amount of outside funding he has received. Over three decades, Welker's research projects have been funded to a total of nearly $2.8 million.

Edgar C. Clausen came to the University of Arkansas as associate professor of chemical engineering in 1981. Clausen received an associate's degree in pre-engineering from Jefferson College, and his B.S.Ch.E. (summa cum laude), M.S.Ch.E., and Ph.D. from the University of Missouri-Rolla. Prior to his arrival at the U of A, he was assistant professor of chemical engineering at Tennessee Technological University. He was named professor of chemical engineering at the U of A in 1985.

Clausen has taught Reactor Design, Advanced Reactor Design, Mass Transport I, Fluid Mechanics, Bioprocess Engineering, Biochemical Engineering, Chemical Engineering Seminar, and Chemical Engineering Laboratory I and II. In addition to his classroom experience, Clausen has served in various capacities for industry. Over the years he has been a process design engineer with Monsanto, summer professor for DuPont, resident consultant for Union Carbide's Nuclear Division, and both vice president for research and vice president for research and development with Bioengineering Resources, Inc. Clausen's professional and honorary affiliations include the AIChE, Tau Beta Pi, Phi Theta Kappa, Phi Kappa Phi, Sigma Xi, and Omega Chi Epsilon. He has also served as a consultant for industrial and governmental groups. Clausen has been granted membership in American Men and Women of Science, Invisible College of Bio-Energy, Outstanding Young Men of America, and Men of Achievement, and he is listed in the *Dictionary of International Biography* and on the membership roles of numerous Who's Whos.

Professor Clausen has authored or jointly authored nineteen books or book chapters. He has published more than 170 articles in professional

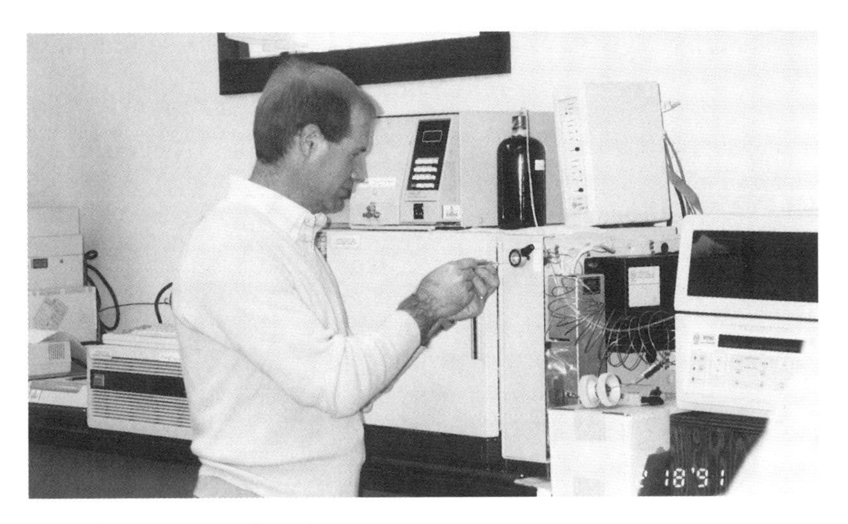

Edgar Clausen. (ChemE collection)

journals or proceedings and has written seventeen university-related research reports. He has participated in more than 180 presentations at scholarly and professional meetings and has given twenty-two invited lectures or seminars. His outside funding for research has topped $4 million.

Clausen's scholarly output has not gone unnoticed. He has won the Halliburton Foundation Award of Excellence for Outstanding Research three times, has been named the Phillips Outstanding Faculty Member for the College of Engineering, and has earned the University of Arkansas Alumni Association's Distinguished Faculty Achievement Award for Research.

William A. Myers was named assistant department head and instructor with the Department of Chemical Engineering in 1984. His homecoming—Myers received his B.S.Ch.E. and M.S.Ch.E. from the U of A—followed a distinguished military career, which included tenures as director of Planning for the Department of Energy's Office of Military Application, assistant for Air Force Nuclear Matters for the Office of the Assistant to the Secretary of Defense for Atomic Energy, chief of the Defense Intelligence Agency's Nuclear Weapons Branch, and research associate at the University of California's Lawrence Livermore National

Laboratory. He retired from the U.S. Air Force on October 1, 1984, with the grade of Colonel.

Myers has taught all three of the Chemical Engineering Laboratory courses. He has also taught Chemical Engineering Design I, Corrosion, and Engineering Materials.

Myers has authored or jointly authored thirty-seven articles in scholarly journals, has made seventeen presentations before scholarly and professional meetings, and has received $147,000 in outside funding to support his research. He has also worked on the U of A's Faculty Senate, the ROTC Programs Committee, the General Education Core Curriculum Committee, and the Student Aid Committee. He is a registered professional engineer in New Mexico. He is a member of the American Society for Metals, the American Chemical Society, and Alpha Chi Sigma. Myers earned the singular distinction of being named Grand Master Alchemist (National President) of Alpha Chi Sigma for the 2000–2002 term.

William A. Myers. (The *Hexagon* of Alpha Chi Sigma, Spring 2000)

Richard K. Ulrich became an assistant professor in the Department of Chemical Engineering in 1987. In 1991, he became an associate professor, and, in 1995, he earned the rank of full professor. Since 1999, Ulrich has had a joint appointment with chemical engineering and electrical engineering. Prior to his arrival in Fayetteville, Ulrich had been a software engineer for Radian Corporation and a research engineer with Texas Instruments. He earned his B.S.Ch.E. from the University of Texas, his M.S.Ch.E. from the University of Illinois, and his Ph.D. from Texas.

Over the years, Ulrich has taught courses in both the chemical

Richard Ulrich (ChemE collection)

engineering and the electrical engineering departments. Among his chemical engineering offerings have been Mass I, Fluid Mechanics, Polymer Science, Materials Science, Microelectronic Materials and Fabrication, and Graduate Seminar. He has supervised thirty graduate students, has served on the committees of forty-seven others, and has guided fourteen undergraduate special projects.

Ulrich has been very active in his service to the university, the college, and the department, serving on various committees, councils, and panels. Outside of the university, he has performed duties as an officer in numerous professional organizations, has reviewed for various journals, programs, and conferences, and has performed consulting work for organizations such as Novatek and Stark Corporation. Ulrich has chaired or co-chaired eleven professional conferences and has organized three symposia. He is a member of the International Electronics and Electrical Engineers, the International Microelectronics and Packaging Society, and the Electrochemical Society.

Ulrich has over sixty scholarly publications to his credit and has authored or jointly authored six books, including *Understanding Mass Transfer,* which is used in the department's mass transfer classes. He has taken part in the production of twenty reports and has made presentations before sixty-nine professional and scholarly organizations. He has received outside funding of more than $7 million to support his research.

In recognition of his dedication to research and teaching, Ulrich has received the Texas Instruments Outstanding Researcher Award, the Halliburton Outstanding Researcher Award, the Phillips Outstanding Faculty Award, and the AIChE Student Chapter Outstanding Lecturer Award.

W. Roy Penney arrived at the Department of Chemical Engineering in 1989 to replace James Gaddy as department head. Penney received his B.S. and M.S. in mechanical engineering from the University of Arkansas and his Ph.D. in chemical engineering from Oklahoma State University. After a brief tenure as chair of the U of A department, Penney relinquished his duties to focus on teaching and research.

W. Roy Penney. (ChemE collection)

Although he had worked as an adjunct professor at Washington University in St. Louis, most of Penney's experience before coming to the U of A had been in industry. Since the early 1960s, he had worked in various management and research positions with companies such as Phillips Petroleum, Monsanto, A. E. Staley, and Henkle Corporation. Even after his appointment at the U of A, Penney has continued to maintain his ties with industry through summer and sabbatical work for S & B Engineers and Dow Corning.

Penney has engaged in numerous scholarly pursuits resulting in thirty-five publications and thirty-one scholarly or professional presentations. He has also presented eleven invited seminars. He holds memberships in the AIChE, Tau Beta Pi, Pi Tau Sigma, and Omega Chi Epsilon, and has been awarded two patents.

Thomas O. Spicer became assistant professor of chemical engineering at the University of Arkansas in 1984. He has since been promoted to associate professor (in 1988) and professor (in 1996). In July 2001 he began service as interim department head. Spicer received his B.S.Ch.E., M.S.Ch.E., and Ph.D. from the University of Arkansas.

Spicer has authored or jointly authored nine articles for professional journals, proceedings for twenty-six symposiums, and twenty-seven research reports. He has also given nine invited lectures and presentations and has made fifteen presentations at scholarly and professional meetings. Funding for his sponsored research projects has topped $3.5 million. He has taught Thermodynamics I, Heat Transport, Computer Methods, Reactor Design, Process Control, Safety, Transport I, Advanced Chemical Engineering Calculations, Advanced Thermodynamics, and Graduate Heat Transfer.

Tom Spicer. (ChemE collection)

Professor Spicer is a member of the Air and Waste Management Association, the AIChE, the American Society of Engineering Education, Omega Chi Epsilon, Sigma Xi, and Tau Beta Pi. He has a long record of community and university service, including work with Habitat for Humanity, the Cooperative Emergency Outreach, the Chemical Engineering Graduate Studies Committee, and the College of Engineering Academic Ethics Board.

Alfred A. Silano came to the University of Arkansas in 1985 as a visiting research professor of chemical engineering. That title became simply research professor when he began working full-time at Fayetteville in 1987. Silano earned a B.S. degree in physical sciences and an Ed.M. in secondary school science education from Rutgers University, an M.S. in engineering science from the Newark College of Engineering, and a Ph.D. in applied mechanics from Rutgers.

Before joining the faculty at the U of A Department of Chemical Engineering, Silano had been a professor in the Department of Chemistry-Physics, Kean College of New Jersey. He had also served as a visiting investigator and consultant for the High Pressure Materials Research Lab of Rutgers' College of Engineering. Silano taught physics at Highland Park High School, Highland Park, New Jersey, from 1953 to 1960.

Alfred A. Silano (ChemE collection)

Professor Silano is a member of the American Association of Physics Teachers, the American Physical Society, Sigma Xi, and the International Association for the Advancement of High Pressure Science and Technology. He has authored or jointly authored eighteen publications, presented his findings at thirteen scholarly or professional meetings, and prepared a twenty-eight-lecture video version of his class on High Polymer Theory and Practice.

At the U of A, Silano designed and developed the Materials Torsion Apparatus (1989–1992), a system that has low temperature capabilities to permit the study of the torsional shear behavior of solid polymers to temperatures of 90K at a variety of shear strain rates and environmental conditions. From 1992 to 2000, Silano designed and developed the Tensile-Torsion Apparatus, a computerized system that evaluates the mechanical behavior of solid polymers under tensile or torsion stress fields, individually or in combination.

Michael D. Ackerson became assistant professor of chemical engineering at the University of Arkansas in 1988. He was promoted to associate professor in 1993. Prior to his appointment to the faculty, Ackerson served as a graduate teaching assistant in chemical engineering at the U of A and as a graduate research assistant with the Biomass Laboratory in the Department of Chemical Engineering, University of Missouri-Rolla.

Ackerson earned his B.S.Ch.E. (1979) and his M.S.Ch.E. (1981) from the University of Missouri-Rolla. He completed his Ph.D. at Arkansas in 1988.

Outside the academic setting, Ackerson has worked as an engineering analyst with Phillips Petroleum in Bartlesville, Oklahoma, and as a summer engineer for the agricultural division of Mobay Chemical Company, Kansas City, Missouri.

Michael D. Ackerson

At the U of A, Ackerson conducts research into biochemical engineering and teaches undergraduate courses in thermodynamics of both single- and multi-component systems. He has developed three graduate-level courses—Estimation of Physical and Thermodynamic Properties,

Advanced Fermentation Studies, and Bioseparations—and has substantially revised the graduate Advanced Thermodynamics course.

Ackerson has authored or co-authored more than fifty articles in refereed and non-refereed scholarly journals and has presented his findings at more than twenty professional seminars, meetings, and symposia. His research has attracted more than $750,000 in outside funding. He holds fifteen U.S. and foreign patents. Ackerson's scholarly achievement has been recognized through numerous awards, including the Ray C. Adam Chair for Young Faculty.

Since 1994 Ackerson has been working to commercialize two technologies developed at the U of A, solvent dewaxing and catalytic hydroprocessing. In April of 2000, a licensing agreement was signed for a $130 million solvent dewaxing plant, the first to use the new dewaxing technology. Less than two years later, in January 2002, a license agreement was signed for the first commercial diesel hydrotreater using the new catalytic hydrotreating technology. The project has been fast-tracked, and the unit will come on stream in August 2002 at a refinery in New Mexico. Ackerson expects that another twenty-five of these hydrotreating units will be licensed and built in the next five years.

Chapter Eight

The Babcock Years, 1990–2001

When Robert Babcock took over as chair of the Department of Chemical Engineering on July 1, 1990, he inherited a department that had achieved unparalleled heights during the previous decade. Yet the growth of the department made running it increasingly complex. The question on everybody's mind: could the achievements of the previous years be continued and expanded? Indeed, they could. The Babcock years, which lasted until he stepped down as department head on July 1, 2001, simply built on the successes of earlier times and increased the department's commitments to instructional and research excellence.

To help handle the complexities of modern university administration, Babcock enlisted Professor Jim Turpin as associate department head for the undergraduate program and Professor Edgar Clausen as associate department head for the graduate program. Colonel William Myers served as assistant department head and director of instructional laboratories.

The early part of the Babcock years brought a revitalized interest in recruiting graduate students to the department. Although its commitment to undergraduate education remained strong, the department actively sought highly qualified candidates for graduate studies. To that end, faculty members hosted informational luncheons for senior-level chemical engineering students, offered seminars at nearby schools with applicable non-B.S.Ch.E. programs, and held Open Houses for prospective graduate students. The graduate studies committee also worked on the recruitment of international students and—reflecting general higher-education trends—developed a successful marketing plan to advertise its merits for potential students. As a result of these efforts, the department experienced dramatic increases in the graduate enrollment, especially for the Ph.D. program, within a few years.

The undergraduate enrollment in chemical engineering continued to evolve during the 1990s and into the twenty-first century. At the beginning of the Babcock era, the average ACT score for incoming freshman had reached 27, and more and more women, minority, and international students called the Department of Chemical Engineering home. These students could look forward to a bright future, according to most observers; one professional magazine, *Graduating Engineer,* noted in January 1991 that chemical engineering ranked among the ten most highly rated occupations in the United States.

Students during the nineties saw the department undergo numerous changes, most notably among the faculty. The decade saw the retirement of faculty mainstays such as Charles Thatcher and Charles Springer, but it also witnessed the hiring of many new, young educators. In 1992, the Department of Chemical Engineering received 10/10 funds from the federal Department of Higher Education for new faculty start-up equipment and to fund five new doctoral assistantships. Besides the new graduate assistants, the money allowed the department to lure dynamic and innovative young teachers and researchers to the faculty. These new faculty members complemented the department's more established researchers and teachers. The chemical engineering program also stayed true to its roots in maintaining close relationships with former students and to bringing those students back as part-time, and even full-time, faculty members, as is evident from the following list of faculty hired during the Babcock era.

1. Carolyne Kincy Garcia. Instructor in Chemical Engineering (1990–1999). B.S., M.S., University of Arkansas.

2. Maria R. Coleman. Assistant Professor of Chemical Engineering (1992–1997); Associate Professor (1997–1998). B.S., Louisiana Tech University; Ph.D., University of Texas.

3. Robert R. Beitle. Assistant Professor of Chemical Engineering (1993–1997); Associate Professor (1997–). B.S.Ch.E., M.S.Ch.E., Ph.D., University of Pittsburgh.

4. Sharon A. Driscoll. Assistant Professor of Chemical Engineering (1993–2000). B.S.Ch.E., University of Washington; M.S.Ch.E., Ph.D., Ohio State University.

5. Gregory J. Thoma. Assistant Professor of Chemical Engineering (1993–). B.S.Ch.E., M.S.Ch.E., University of Arkansas; Ph.D., Louisiana State University.

6. John Francis Bushkuhl. Visiting Assistant Professor of Chemical Engineering (1994–). B.S.Ch.E., University of Arkansas.

7. Valentin E. Popov. Visiting Professor of Chemical Engineering (1994–1997). M.S., Moscow State University; Ph.D., Agro-Physical Institute.

8. Robert A. Cross. Research Professor of Chemical Engineering (1995–). B.S.Ch.E., University of Arkansas; M.S., Massachusetts Institute of Technology.

9. Heather Logue Walker. Research Assistant Professor (1996–1999). B.S.Ch.E., M.S.Ch.E., Ph.D., University of Arkansas.

The first year of Babcock's chairmanship looked remarkably similar to Gaddy's last, in terms of the chemical engineering curriculum. For the 1990–1991 school year, students would have signed up for the following:

Freshman Year

Fall	Spring
MATH 2555, Calculus I	MATH 2565, Calculus II
CHEM 1123, University Chemistry	CHEG 1123, Intro. to Chem. Engineering II
CHEM 1131, University Chemistry Lab	ENGL 1023, Technical Composition
ENGL 1013, Composition	Humanistic-Social Sciences Elective
CHEG 1113, Intro. to Chemical Engineering I	CHEG 1212, Chemical Engineering Lab I

Sophomore Year

Fall	Spring
MATH 2573, Calculus III	MATH 3403, Differential Equations
CHEM 3604, Organic Chemistry	CHEM 3614, Organic Chemistry
PHYS 2053, University Physics I	PHYS 2073, University Physics II
PHYS 2061, University Physics Lab I	PHYS 2081, University Physics Lab II
CHEG 2133, Momentum Transport	CHEG 2313, Thermo. of Single-Comp. Sys.
Humanistic-Social Sciences Elective	Humanistic-Social Sciences Elective

Junior Year

Fall	Spring
CHEM 3504, Physical Chemistry I	CHEM 3514, Physical Chemistry II
MEEG 2003, Statics	MEEG 3013, Mechanics of Materials
CHEG 3143, Heat Transport	CHEG 3333, Chem. Engr. Reactor Design
CHEG 3322, Chemical Engineering Lab II	CHEG 3153, Mass Transport I
CHEG 3323, Thermo. of Multi-Component Systems	ECON 2013, Principles of Economics
Humanistic-Social Sciences Elective	

Senior Year

Fall	Spring
CHEG 4163, Mass Transport II	CHEG 4332, Chemical Engineering Lab III
CHEG 4413, Design I	CHEG 4443, Design II
CHEG 4423, Automatic Process Control	ELEG 3903, Electric Circuits
Technical Electives	Technical Electives
CHEG 3221, Chemical Engineering Seminar	Humanistic-Social Sciences Elective

Chemical engineering students could choose from one of seven technical options to fulfill their technical electives.

The next year, the department added two organic chemistry labs to complement the Organic classes for the sophomore years plus CHEG 3253, Computer Methods, to the required coursework. Two years later, in 1993, History of the American People to 1877 became a first-semester requirement, replacing a humanities/social science elective from the sophomore year, and Basic Economics—Theory and Practice replaced Principles of Economics. For the 1996–1997 school year, the department stipulated that freshmen had to take three hours of foreign language in place of a humanities/social science elective; the foreign language requirement was dropped by 1997.

By the 1996–1997 school year, a new class had been added to the chemical engineering course list: CHEG 4263, Environmental Experimental Methodology, taught by Greg Thoma as an introduction "to experimental design, environmental analytical method quality assurance of analytical measurements, sample collection and preservation"; the class included "laboratory work necessary to support a field scale tracer experiment." Two years later, Professor Cross began guiding students through CHEG 5013, Membrane Separation and System Design, which looked into the "theory and system design of cross flow membrane process—reverse osmosis, nanofiltration, ultrafiltration, and microfiltration—and applications for pollution control, water treatment, food and pharmaceutical processing." That same year, Chemical Process Safety became a required senior-year course.

By the end of the Babcock era, the chemical engineering curriculum looked like this:

Freshman Year

Fall	Spring
MATH 2554, Calculus I	MATH 2564, Calculus II
CHEM 1123, University Chemistry II	CHEG 1123, Intro. to Chem. Engineering II
CHEM 1121L, University Chemistry II Lab	ENGL 1023, Technical Composition II
ENGL 1013, Composition I	Humanistic-Social Sciences Elective
CHEG 1113, Intro. to Chemical Engr. I	CHEG 1212L, Chemical Engineering Lab I
HIST 2003, Hist./American People to 1877	Humanities/Social Sciences Elective

Sophomore Year

Fall	Spring
MATH 2574, Calculus III	MATH 3404, Differential Equations
CHEM 3603, Organic Chemistry I	CHEM 3613, Organic Chemistry II
CHEM 3601L, Organic Chemistry I Lab	CHEM 3611L, Organic Chemistry II Lab
PHYS 2054, University Physics I	PHYS 2074, University Physics II
PHYS 2050L, University Physics Lab I	PHYS 2070L, University Physics Lab II
CHEG 2313, Thermo. of Single-Comp. System	CHEG 3323, Thermo. of Multi-Comp. Sys.
CHEG 3221, Chemical Engineering Seminar	CHEG 2133, Momentum Transport

Junior Year

Fall	Spring
Chemistry Elective	Chemistry Elective
MEEG 2003, Statics	MEEG 3013, Mechanics of Materials

CHEG 3143, Heat Transport	CHEG 3333, Chem. Engr. Reactor Design
CHEG 3232L, Chemical Engineering Lab II	CHEG 3153, Mass Transport I
CHEG 3253, Chem. Engr. Computer Methods	ECON 2143, Basic Economics
Humanities/Social Sciences Elective	

Senior Year

Fall	**Spring**
CHEG 4163, Mass Transport II	CHEG 4332L, Chemical Engr. Lab III
CHEG 4413, Chemical Engineering Design I	CHEG 4443, Chem. Engr. Design II
CHEG 4423, Automatic Process Control	ELEG 3903, Electric Circuits
CHEG 4813, Chemical Process Safety	Technical Electives
Humanities/Social Sciences Elective	CHEG 4423, Process Control
	Humanities/Social Sciences Elective

For the chemical engineering profession as a whole, the 1990s witnessed an even greater blurring of already fuzzy distinctions between industrial and educational research. University faculty members around the nation took more and more leaves for temporary industrial research and development positions and to make contacts that would be important for their own and for their students' futures. University faculty members created their own tech companies to exploit their years of hard work and their expertise within their areas of specialty. At the same time, chemical engineers who had spent most of their careers in the private sector began filling temporary, and sometimes permanent, positions at universities and colleges. All of these trends, and questons they raised about the fundamental purpose of university-based research, were evident at the University of Arkansas during the Babcock years.

At the beginning of Dr. Babcock's tenure, the Department of Chemical Engineering was engaged in research funded at $5 million. The level of funding remained high throughout the nineties.

Professor Jerry Havens performed some of the most noted research within the department with his internationally renowned studies of heavy gas dispersion. When the nineties began, Havens was busily constructing a wind tunnel "designed expressly for conducting dispersion experiments to obtain development and validation of mathematical models for dispersion of accidentally released hazardous gases." Havens used the wind tunnel—the world's largest of its kind at 120 feet long by 20 feet wide and

Jerry Havens's Wind Tunnel. (University Relations collection)

containing two 6-foot wide, 75 horsepower fans—to validate computer-generated predictions of how heavy gases disperse in light winds. The wind tunnel allowed for the repetition and fine-tuning of experiments at less expense than actual field tests. Havens's research has been vital to both the University of Arkansas and to governmental and private agencies interested in understanding heavy gas dispersal.

Besides Havens's research, other faculty on campus expanded the chemical engineering research horizons. Research Professor A. A. Silano designed a torsion apparatus to look into temperature-dependent shear deformation of polymers. His work helped predict and explain such behaviors as fatigue, creep, stress, finite strain, stress cracking, and the thermal expansivity of plastics. Michael Ackerson studied tissue culture and dewaxing tall crude oils. In 1995, Maria Coleman received a five-year Presidential

Jerry Havens, distinguished professor of chemical engineering, and Dr. Heather Walker, manager of the Chemical Hazards Research Center wind tunnel. (University Relations collection)

Faculty Fellowship funded by the National Science Foundation for $500,000; Coleman used the fellowship to research and develop polymers for use as gas separation membranes to recover oxygen in such places as hospitals and to purify air in spacecraft.

Although graduate students assisted faculty members and performed their own research projects, a relatively unique facet of the University of Arkansas's Department of Chemical Engineering was, and still is, the emphasis placed on undergraduate research. Besides taking required research courses, undergraduates benefited from the leadership of faculty members willing to develop and refine undergraduate research opportunities. Greg Thoma began offering the nation's only field-scale laboratory for undergraduates in the mid-nineties for his course, Introduction to Experimental Environmental Engineering; typical research involved running dye trace studies in the White River to understand the safety limits for chemical dumping.

Robert Beitle, along with Ingrid Fritsch of the chemistry department, served as co-coordinator of a three-year interdisciplinary Research Experience for Undergraduates program. Funded with a National Science Foundation grant, the Research Experience focused on separation technologies and included student participants from the fields of chemical engineering, chemistry, food science, biochemistry, and physics. Ten students, selected from a pool of nationwide applicants, engaged in research over summer holidays.

Research activities require space, and, although the chemical engineering department did not benefit from any major construction projects during the Babcock years, it did upgrade its facilities. Although the Bell Engineering Center was one of the newest buildings on campus, the nature of chemical engineering research and instruction required a constant search for better ways of conducting both.

A student computer laboratory opened for the 1994–1995 school year. The lab included computational programs such as Math CAD, TK Solver for Windows, Maple, and Hysim, and it granted student access to word-processing and spreadsheet software. The department also created new Multi-Media and Media Development Laboratories to increase the instructional efficiency of staff members; a Student Media Development Lab gave students valuable tools for in-class presentations.

Engineering Hall received a facelift in the early nineties. The renovation project included the installation of tinted, insulated windows and a new roof. The air-conditioning, telephone, and electrical systems were updated, and data circuits were installed for computer networking. To address safety concerns, workers improved handicap access to the building and installed new fire doors and alarms. After its renovation, Engineering Hall became the permanent location of all or parts of the departments of electrical engineering, biological and agricultural engineering, and computer science engineering; once these departments were settled in, all engineering departments were housed along Dickson Street, within a few blocks of one another.

When they weren't taking classes in those buildings along Dickson, chemical engineering students of the nineties continued to participate in the activities that earlier students had, including the more subdued Engine Week and extracurricular activities such as the AIChE, Alpha Chi Sigma, and Omega Chi Epsilon. The AIChE hosted an annual golf tournament that proved to be very popular; cookouts in the fall and spring also brought many participants—one student recalled that "faculty members would make chili and students would eat it and vote on the best recipe." But for many of the program's students, the rigors of the chemical engineering curriculum precluded much outside activity. Indeed, by the Babcock era chemical engineering students developed activities that combined academic and social life.

The student lounge in Bell Engineering Center became the focal point for many students' outside-the-classroom experience. Students congregated there to prepare homework, study for exams, enjoy one another's company, relax, and eat. Late-night and early-morning study sessions blurred the distinction between leisure and work, and chemical engineering students of the period invariably look back on the two as inseparable. Card games became commonplace, remembered one student of the era: "Another thing we did for fun around the department when I was a grad student there in the early '90s was play a lot of Spades. We had tournaments, although I don't remember what, if any, prizes were ever awarded."

Chemical engineering students had always stuck together. The nature of the chemical engineering program did not allow for excessive amounts of extracurricular activity; too much partying, too much "hanging out," and the student would not be able to handle the difficulties of the chemical engineering coursework. Instead, the schoolwork fostered a camaraderie among the students, a "we're all in this together" sort of feeling. Unlike some academic majors that allow for liberal doses of elective coursework, chemical engineering majors ended up in many of the same required classes with the same people, and they thus got to know one another quite well by the end of their freshmen or sophomore years. Such comments as "there was a groups of us, when I was an undergrad . . . [and] we did things together" are typical and explain why the campus life of a chemical engineering student might have been much different from that of students in other fields. Even off campus, chemical engineering students tended to stick together, gathering at apartments or houses for parties and holding cookouts.

Chemical engineering students judge the Annual Faculty Chili Cook-off in the department's student lounge. (ChemE collection)

The department also boasted a student barbershop quartet that, in the words of a one-time member, "sang for its own amusement and also performed at functions like the annual department banquet and Christmas parties around Bell Engineering Center." A group of chemical engineering students also created a brass ensemble that played in local churches from time to time.

The Department of Chemical Engineering under Professor Babcock reached out to new and old friends during the 1990s. In January of 1992, Babcock went on a goodwill tour of industry in the gulf coast region. The tour established contacts that would prove useful for student and graduate employment in addition to providing new sources of potential financial support. A year earlier, Arkansas Chemical Engineering Supporters (ARChES) had been created to aid the department in student recruiting, industrial relations, and alumni support. In September 1992, the department sponsored a Second Alumni Reunion for former students who graduated between 1902 and 1960; that same year plans began to be set in motion toward a "major event" to celebrate the chemical engineering program's centennial in 2002.

As of the writing of this book, Professor Babcock's successor as department head has not been named; Tom Spicer has been interim chair since July 1, 2001. Although uncertainty remains as to who will ultimately replace Babcock on a permanent basis, the temporary transition period is quite normal for the Department of Chemical Engineering. One thing is certain, though: whoever is chosen will have large shoes to fill.

Biographical Sketches, 1990–2001

Maria Coleman arrived at the University of Arkansas as assistant professor of chemical engineering in July of 1992. Coleman, the department's first female faculty member, had received her B.S.Ch.E. from Louisiana Tech University and her Ph.D. from the University of Texas. She received a promotion to the rank of associate professor in 1997. In August of 1998, Coleman left the U of A for a position with the Department of Chemical and Environmental Engineering at the University of Toledo.

While at the U of A, Coleman published numerous articles in venues such as *Journal of Polymer Science, Macromolecules,* and *International Journal of Biochromatography,* and made twenty presentations at professional or scholarly meetings. The articles and presentations most often reflected her research interest in the area of the development of polymeric materials for membrane-based separations applications for gas separations. She received about three hundred thousand dollars in outside funding for research and teaching development while at the U of A and was awarded a five-year Presidential Faculty Fellowship by the National Science Foundation in 1995. Dr. Coleman taught graduate and undergraduate classes on separations, fluids, transport, and polymers.

Maria Coleman. (University Relations collection)

Coleman is a member of the North American Membrane Society, the AIChE, the ACS, Sigma Xi, Omega Chi Epsilon, and Tau Beta Pi. Among her honors at Arkansas, she was awarded the Halliburton Outstanding Teaching Award, the Ray C. Adam Young Faculty Chair, and the Phillips Petroleum Company Outstanding Faculty Member of the College of Engineering.

Robert Beitle arrived at the Department of Chemical Engineering in 1993 as an assistant professor. Beitle received his B.S., M.S., and Ph.D. from the University of Pittsburgh, where he served as a graduate research and teaching assistant from 1987 to 1993. He is a registered professional engineer with the state of Arkansas.

Beitle has been awarded more than $440,000 in funding for research on biochemical engineering, bioseparation, and microbial waste treatment. He has published his findings in sixteen scholarly and professional journals, and he has made twenty-four presentations at professional meetings. Beitle has also delivered seven talks by invitation. He has served as editor of *BIOTechnologist* and the *American Chemical Society BIOT Division Newsletter* and has been a session chair for the American Chemical Society and the American Institute of Chemical Engineers; his work as a reviewer has been utilized by the National Science Foundation, *Biotechnology and Bioengineering, Biotechnology Progress,* and *Separation Science and Technology.* He is a member of the AIChE, the ACS, the North American Membrane Society, Tau Beta Pi, and Omega Chi Epsilon.

Robert Beitle. (ChemE collection)

Doctor Beitle received the title of associate professor in August 1998. His course offerings include the undergraduate classes Chemical Engineering Reactor Design and Chemical Engineering Design I and the graduate Biochemical Engineering and Chemical Engineering Numerical Methods. Beitle's excellence in teaching and research has resulted in his winning the Halliburton Outstanding Research Award (1995), the Texas Instruments Research Award (1997), and the Chemical Engineering Teaching Award (1998). He has also served as chemical engineering representative for the ABET EC 2000 team and has been faculty advisor for both Tau Beta Pi and Omega Chi Epsilon. His service to the university has included memberships on the Department of Chemical Engineering Graduate Studies Committee, the College of Engineering Academic Programs Committee, the College of Engineering Teaching Innovation Committee, the College of Engineering Curriculum Committee, the University of Arkansas Biosafety Committee, and the Winthrop Rockefeller Distinguished Lectureship Committee. Beitle has advised six undergraduates, twelve master's students, and one doctoral student.

Sharon Driscoll became assistant professor in the Department of Chemical Engineering in 1993. Driscoll received her B.S.Ch.E. from the

University of Washington and her M.S.Ch.E. and Ph.D. from the Ohio State University. Her previous academic experience included stints as graduate research assistant, graduate teaching assistant, and academic advisor at the Ohio State University and as a technician with the University of Washington.

Driscoll holds memberships in Phi Kappa Phi, the AIChE, the Association of Women in Science, and the American Society for Engineering Education. She is a licensed professional engineer in Arkansas. She has presented the results of her research activities in twenty-seven published articles and proceedings. She has also given presentations at seventeen professional meetings and has accepted sixteen invitations to talk at seminars and other forums. Driscoll has been a proposal reviewer for the Department of Education, the National Science Foundation, and the U.S. Civilian Research and Development Foundation; she also has reviewed papers for the *Journal of Physical Chemistry* and the American Chemical Society.

While at the University of Arkansas, Driscoll taught Introduction to Chemical Engineering I, Mass Transport I, Reactor Design, Chemical Engineering Laboratory III, Advanced Reactor Design, Advanced Equilibrium Stage Separations, and Graduate Seminar. She supervised eighteen undergraduate and four graduate students in their research projects. Reflecting her excellent teaching skills, Driscoll was awarded the Ray C. Adam Chair for Young Faculty and the Texas Instruments Award for Outstanding Service to Students in 1998. She served on one university-level committee, two college committees, and seven departmental committees.

Doctor Driscoll left the department in August 2000.

Gregory Thoma joined the chemical engineering faculty at the University of Arkansas in 1993. His appointment as assistant professor marked a return to his alma mater; he earned both his B.S.Ch.E. and M.S.Ch.E. from the U of A. Thoma received his Ph.D. in 1994 from Louisiana State University, where he wrote his dissertation, "Studies on the Diffusive Transport of Hydrophobic Organic Chemicals in Bed Sediments," under former U of A professor Louis Thibodeaux. His dissertation was named "best dissertation" for 1994 by the Baton Rouge/New Orleans Division of AIChE.

Thoma has had articles in sixteen peer-reviewed publications and has made presentations to thirty-one scholarly and professional meetings. He has also published ten reports or proceedings. His research has been funded through numerous grants to more than four hundred thousand dollars. He is a registered professional engineer and has performed consulting work for ALCOA, McClelland Consulting Engineers, and for Ascension Parish, Louisiana. He has performed volunteer work for the Peace Corps, Neighbor Helping Neighbor, the Red Cross, Ozark Mountain Rescue, and Mother Theresa's Home for the Dying and Destitute.

Greg Thoma. (ChemE collection)

Thoma was named associate professor of chemical engineering in 1999. Over the years he has taught Heat Transfer, Advanced Heat Transfer, Chemical Engineering Laboratory II, Design II, Computer Methods, Air Pollution Control, and Experimental Environmental Methodology, and he has supervised the research of seven undergraduate and four graduate students. His service to the university has included participation in the College of Engineering Environmental Steering Committee, the College of Engineering Service Course Committee, the U of A Toxic Substances Committee, the U of A Environmental Task Force Committee, and the Chemical Engineering Graduate Committee Admissions Committee. He has been named the Texas Instruments Outstanding Researcher in Chemical Engineering and has twice been honored as the Texas Instruments Outstanding Teacher in Chemical Engineering.

Robert A. Cross returned to his undergraduate alma mater as research professor of chemical engineering and co-director of the Center for Membrane Applied Science and Technology in 1995. Cross earned his

B.S.Ch.E. from the University of Arkansas in 1957 and his M.S.Ch.E. from the Massachusetts Institute of Technology in 1959.

Prior to his return to the U of A, Cross had a productive career in private industry. From 1963 to 1972, he served as senior project engineer, project manager, manager of film products, assistant director of research, and director of process research with Amicon Corporation. He was president of Romicon, Inc., from 1972 to 1988, and CEO, chairman, and technical director of Bioken Separations, Inc., from 1989 to 1991. His last stop before returning to the chemical engineering department was as vice president, general manager, and technical director of the Separations Systems Division of Cuno, Inc., from 1991 to 1994.

Robert A. Cross. (ChemE collection)

At the U of A, Cross teaches Membrane Technology and Chemical Engineering Process Design courses, and he has set up the Membrane Separations Center to coordinate membrane-related research and to serve as a focal point for obtaining government and corporate support. His research, before and after his return to the U of A, has resulted in thirty publications and presentations, and he holds five U.S. patents.

Epilogue: 2002 and Beyond

Entering its second century, the University of Arkansas Department of Chemical Engineering continues to seek new challenges and to broaden its horizons. Perhaps the most telling signs of the department's future growth and development are the international partnerships it is forging with educational institutions around the world.

In early 2000, Brian Freedman, director of the International Development Office of the University of Newcastle, Australia, contacted the University of Arkansas with hopes of establishing department-to-department connections. Newcastle had been actively seeking these inter-departmental connections with high-quality schools in the United States and around the world in an effort to counteract many Australian students' reluctance to travel abroad. Freedman traveled to the U of A, where he met with Assistant Department Head William Myers, and, with the

U of A Chemical Engineering Students attending the University of Newcastle in Australia in 2001, *l. to r.*, Chris von der Mehden, Laney Philpott, Chris Barnes, James B. Butler. Missing from the picture are U of A Chemical Engineering students Noel Romey and Christy M. White, who also studied at Newcastle during the July to November term. (Courtesy of Christy White)

A 4 EL DEBER ▶ SANTA CRUZ

Santa Cruz de la Sierra
Domingo 1 de octubre de 2000

La U de Arkansas tiene la fórmula para perderle el miedo a química

TECNOLOGÍA. *Los jóvenes cruceños aún evitan carreras técnicas pese a ventajas*

JUAN CARLOS RIVERO

En medio del tumulto de gente de Expocruz, la cabeza blanca del Dr. William Myers se abrió paso para llegar hasta el sector agrícola. Miró las bolsas de abono y expresó con asombro: "¡Este abono es hecho en Argentina! Increíble que un país como Bolivia, repleto de materia prima para fabricarlo, tenga que comprarlo de su vecino".

De esta manera, el Dr. Myers confirmó su teoría de que Bolivia necesita de ingenieros químicos para transformar su riqueza en productos terminados.

Siendo él catedrático de esa carrera en la Universidad de Arkansas, se dio cuenta que del numeroso grupo de jóvenes cruceños que estudia allí, ninguno se sienta en sus aulas. La mayoría se inclina por las ramas de negocios y humanidades. ¿Miedo a los números y a la tecnología?

Es posible. Pero precisamente el Dr. Myers llegó a Santa Cruz para disipar temores, para convencer a jóvenes postulantes de que pueden aprender química en el ambiente casi familiar de la Universidad de Arkansas.

"Exigimos mucho a nuestros estudiantes", dice, "pero basta con que alguien rinda al máximo de sus habilidades para que alcance el éxito".

Acompañando al Dr. Myers, vino también la directora del Centro de Idiomas de esa universidad norteamericana, Dra. Vicki Bergman-Lanier, quien resaltó el tremendo sistema de apoyo con que cuentan.

Asegura que el estudiante internacional no debe sentirse intimidado porque cuentan con centros asistenciales para ayudarlo en sus problemas académicos, diferencias culturales y hasta conflictos emocionales.

Los nombres de todos los egresados desde 1876 están grabados en corredores

La Dra. Bergman considera que el sistema de estudio ofrece gran flexibilidad a la hora de elegir una carrera, porque los primeros dos años el estudiante tiene la libertad de elegir materias de entre 78 disciplinas.

La Universidad de Arkansas favorece al estudiantado cruceño y nacional gracias a un convenio que tiene con la organización Compañeros de las Américas. Esto permite que los bolivianos paguen la misma matrícula que el ciudadano norteamericano que reside en el estado de Arkansas.

CRUCEÑOS. *Los estudiantes de Santa Cruz ganan presencia en la Universidad de Arkansas. Se sienten como en casa*

VISITA. *Myers y Bergman en EL DEBER. Quieren formar jóvenes valores*

UN VISTAZO A LA UNIVERSIDAD DE ARKANSAS

- **Caras conocidas.** Cerca de un centenar de bolivianos forma parte de los 15.000 estudiantes.
- **Un universo de opciones.** Ofrecen 87 títulos en 78 carreras, sin contar el postgrado.
- **Te conocen el nombre.** La mayoría de las aulas tiene 29 o menos alumnos.
- **Arkansas-Bolivia, como si fueran la misma cosa.** El boliviano paga matrícula como uno del lugar. Esto significa un ahorro de hasta 5.000 dólares al año.
- **Todo un pueblo universitario.** El campus se encuentra ubicado en Fayetteville, una ciudad de 55.000 habitantes.
- **Codéate con los mejores:** Los programas de ingeniería están considerados entre los mejores de Estados Unidos.
- **Experiencia:** La fundaron en 1871.
- **Si te interesa:** Director of International Admissions, Lynn Mosesso; e-mail: mosesso@uark.edu; Telf. (501)-575-6246.

¡Adelante puercos filo 'e navaja!

El título de esta nota no es un insulto. Es el grito que hace vibrar a todo el estado de Arkansas cuando su universidad compite en deportes contra sus similares.

Se trata de los Razorbacks o puercos salvajes cuya espalda erizada da la impresión de tener el filo de una navaja. Antes de adoptar el nombre de este tenaz animal, el equipo de fútbol americano iba acompañado de otra mascota un tanto más inofensiva, el cardenal.

Hasta que en 1909, el equipo trituró a su clásico rival, lo que inspiró al entrenador a calificar a sus jugadores como Razorbacks o "puercos filo e' navaja" (a lo camba).

Un partido de los Razorbacks equivale a un clásico Oriente-Blooming multiplicado por tres, donde todos en las tribunas visten rojo y blanco, todos identificados por su alma mater.

Colonel Myers visits Bolivia to recruit chemical engineering students. (*El Deber*, Santa Cruz, Bolivia)

concurrence of Department Head Robert Babcock, set in motion plans for a partnership between the chemical engineering departments of Newcastle and Arkansas. The U of A Fulbright Institute handled the paperwork, and within six months the two universities had drafted and signed a formal agreement.

In May 2000, Eric Kennedy, a chemical engineer from the University of Newcastle, visited the U of A and met with chemical engineering students. One of those students, Noel Romey, who has family connections in Australia, took the lead in pursuing the possibility of studying at

Newcastle. During the summer of 2001, the first group of Arkansas chemical engineers—Romey, Chris Barnes, James Butler, Laney Philpott, Chris von der Mehden, and Christy White—traveled to Australia for the July to November school term. Students from Newcastle are expected to arrive at Arkansas in January 2002 for the spring term.

Another international outreach program began when, in the spring of 2000, Colonel Myers discovered that although almost one hundred Bolivian students called the U of A home, none of them studied chemical engineering. He found this especially curious since Bolivia's chemical industries are underdeveloped and the need for trained chemical engineers there is acute. With the encouragement of Professor Babcock, Myers contacted Max Frydman, a 1986 graduate of the department and a Bolivian citizen, for help in setting up a recruiting trip to Bolivia. Frydman was uniquely qualified to help—through his position as director of Profesores Asociades in Santa Cruz, Bolivia's second-largest city, he counseled Bolivian students who wish to study in the United States, and he was heavily involved with a group known as Partners of the Americas, a group of influential Santa Cruz residents who have sent family members to study at the U of A and other U.S. colleges.

In September 2002, Myers, his wife, Satsuki, and Dr. Vicki Bergman-Lanier, director of the U of A Spring International Language Center, traveled to Bolivia, where they met with Frydman and interviewed prospective students. In the course of three days, the group met with over one hundred students at La Paz, Bolivia's capital; they then traveled to Santa Cruz, where a similar number of interested students turned out. The department is hopeful that the seed planted by Colonel Myers on this trip will soon bear fruit.

The Department of Chemical Engineering's international outreach reflects the globalization of chemical industries and marks a deliberate effort to reach new students. Such exchange programs allow even more students from around the globe to experience the unique balance of research and teaching of the Department of Chemical Engineering and give Arkansas students the chance to learn about their field in different cultural and social environments. The globalization of the chemical industries, and thus of the U of A Department of Chemical Engineering, will only increase over the next hundred years.

EPILOGUE

In one century, the University of Arkansas's chemical engineering program grew from what was only a footnote on campus—really only an idea in the mind of Professor A. M. Muckenfuss in 1902—to one of the university's leaders in research and education. In its first seventeen years of existence, the program graduated only three chemical engineers; those early graduates were little more than chemists who possessed at least some familiarity with civil, mechanical, and electrical engineering. By the end of its first century, the program annually enrolled hundreds of undergraduate and dozens of graduate students.

To compare the facilities of the early twentieth century to those of the early twenty-first staggers the mind. From the early Science Hall to the original Chemistry Building, the old Engineering Building, Engineering Hall, the 1930s Chemistry Building, the Barn, the Engineering Hall Annex, and finally to the present chemical engineering home at Bell Engineering Center and Engineering South, the program has grown by leaps and bounds, all in an effort to increase teaching efficiency and to stay on the cutting edge of technological and research advances.

The people of the chemical engineering program have changed as well. The faculty has evolved from one comprised solely of chemists to one that is well trained in teaching and researching chemical engineering specialties. Even more significant, the students have changed. Notably, as the century wore on, more and more women enrolled in the curriculum; later, minority and international students became common within the program.

It is the people, after all, who make any educational endeavor what it is: a means of passing knowledge on and of increasing understanding. And that is what has made the Department of Chemical Engineering so special. Students who learn from the department's outstanding teachers and researchers go out into the world prepared to carry on the endeavor. There is no reason to doubt that this mission will only continue—indeed, that it will increase—over the next hundred years.

Sources

A footnoted copy of this book is available at the Department of Chemical Engineering, Bell 3202, University of Arkansas, Fayetteville.

Books, Articles, and Chapters

Brown, Kent R. *Fayetteville: A Pictorial History.* Norfolk, Va.: Donning Company/Publishers, 1982.

Campbell, William S. *One Hundred Years of Fayetteville, 1828–1928.* Fayetteville, Ark.: William S. Campbell, 1928.

Colton, Joel, and Stuart Bruchey, eds. *Technology, the Economy, and Society: The American Experience.* New York: Columbia University Press, 1987.

Grayson, Lawrence P. *The Making of an Engineer: An Illustrated History of Engineering Education in the United States and Canada.* New York: John Wiley & Sons, 1993.

Hale, Harrison. *University of Arkansas, 1871–1948.* Fayetteville: University of Arkansas Alumni Association, 1948.

Oxford, Charles. "A History of Chemical Engineering at the University of Arkansas," in *Proceedings of the Symposium: Development and Future of the Chemical Industry in Arkansas.* Fayetteville: University of Arkansas Department of Chemical Engineering, 1983.

Pigford, Robert L. "Chemical Technology: The Past 100 Years." *Chemical & Engineering News* (April 6, 1976): 190–203.

Reynolds, John Hugh, and David Yancey Thomas. *History of the University of Arkansas.* Fayetteville: University of Arkansas, 1910.

Rudolph, Frederick. *The American College and University, A History.* New York: Knopf, 1962.

Simpson, Ethel C. *Image and Reflection: A Pictorial History of the University of Arkansas.* Fayetteville: University of Arkansas Press, 1990.

Westwater, J. W. "The Beginnings of Chemical Engineering Education in the USA." In William F. Furter, ed., *History of Chemical Engineering: Based on a Symposium Cosponsored by the ACS Divisions of History and Industrial and Engineering Chemistry at the ACS/CSJ Chemical Congress, Honolulu, Hawaii, April 2–6, 1979.* Washington, D.C.: American Chemical Society, 1980.

SOURCES

Journals, Magazines, and Newspapers

Arkansas: The Magazine of the Arkansas Alumni Association, Inc.
Arkansas Alumnus
Arkansas Engineer
Fayetteville Weekly Democrat
Keeping Up . . . : A News Publication for University of Arkansas Chemical Engineering Graduates and Friends
Morning News of Northwest Arkansas
University

University of Arkansas Materials

Annual Reports
Cardinal (yearbooks)
Engineering Catalogs
Engineering Bulletins
Razorback (yearbooks)
Reports of the President
University of Arkansas Catalogs
University of Arkansas Bulletins

Interviews

Bell, Gaston
Brown, Robert
Middleton, John P.
Nelms, David
Oxford, Charles
Thibodeaux, Louis J.

Files of the Department of Chemical Engineering

"A Petition to Alpha Chi Sigma from Gamma Chi, University of Arkansas" (1921) photocopy.
"Alfred A. Silano," resumé.
"Bio-Charles W. Oxford," information sheet.
Chemical Engineering Alumni Reunion, 1902–1959, April 1988.

SOURCES

"Class of 1932—Biographical Resumé," copy of file from North Carolina State Alumni Association.
"Comments on 'The Bocquet Year' from Robin G. Moore," handwritten notes.
"Edgar C. Clausen," resumé.
F. Parker Helms to Michael S. Martin, letter, July 7, 2001.
"Gregory John Thoma," resumé.
"J. Reed Welker," resumé.
"James Couper," information sheet.
Jim Kilby to Michael S. Martin, email correspondence, August 1, 2001.
"Jim L. Turpin," resumé.
Jim Stice to Michael S. Martin, email correspondence, September 3, 2001.
John M. Meazle to Michael S. Martin, email correspondence, July 1, 2000.
Joseph R. Jacobs to Michael S. Martin, letter, July 25, 2001.
"Louis Joseph Thibodeaux," resumé.
"M. E. Barker Scholarship" file.
"Maria Rosario Coleman," resumé.
Mark A. Myers to Michael S. Martin, email correspondence, July 23, 2001.
News Release, Naval Research Laboratory, June 9, 1993.
Printout of chemical engineering alumni from University of Arkansas Alumni Association.
J. Reed Welker to Michael S. Martin, email communication, August 9, 2001.
"Resume of Dr. James L. Gaddy."
"Richard K. Ulrich," resumé.
"Robert A. Cross," resumé.
"Robert R. Beitle, Jr.," resumé.
Robert MacCallum to Michael S. Martin, email communication, September 20, 2001.
"Sharon A. Driscoll," resumé.
T. A. Walton to Michael Martin, email correspondence, August 26, 2000.
"William A. Myers," resumé.
"Thomas O. Spicer III," resumé.
William A. Myers to Michael S. Martin, email correspondence, August 9, 2001.
"William Roy Penney," resumé.

Index

B

INDEX